# Mass Spectrometry - Applications and Recent Advances

*Edited by Nahide Gülşah Deniz*

Published in London, United Kingdom

Mass Spectrometry - Applications and Recent Advances
http://dx.doi.org/10.5772/intechopen.1009079
Edited by Nahide Gülşah Deniz

Contributors
Alanood Almansoori, Andriamanarivosoa Rija Razafintsalama, Antonio Randriampahasoavana, Bhaskar Karubothula, Dilek Salkım İşlek, Dnyaneshwar Shinde, Eda Kiriş, Elando Fréda Zamanileha, Gilles Bedoux, Grzegorz Brudecki, Marnella Quetty Lema, Maryam AlQaydi, Miyang Li, Parth Gupta, Picardino Frienduc Vaonalamihanta, Pierre Hervé Ravelonandro, Raghu Tadala, Robert Owen Bussey III, Samara Bin Salem, Stanislas Fulgence Lahadson, Wael Elamin



First published in London, United Kingdom, 2026 by IntechOpen
IntechOpen is the global imprint of INTECHOPEN LIMITED, registered in England and Wales, registration number: 11086078, 167-169 Great Portland Street, London, W1W 5PF, United Kingdom

For EU product safety concerns: IN TECH d.o.o., Prolaz Marije Krucifikse Kozulić 3, 51000 Rijeka, Croatia, info@intechopen.com or visit our website at intechopen.com.

British Library Cataloguing-in-Publication Data
A catalogue record for this book is available from the British Library

Mass Spectrometry - Applications and Recent Advances
Edited by Nahide Gülşah Deniz
p. cm.
Print ISBN 978-1-83634-201-4
Online ISBN 978-1-83634-200-7
eBook (PDF) ISBN 978-1-83634-454-4

# Meet the editor

Prof. Dr. Nahide Gülşah Deniz is a specialist in the design, synthesis, characterization, and investigation of novel bioactive organic compounds. Dr. Nahide Gülşah Deniz has been a Professor in the Division of Organic Chemistry, Department of Chemistry, Engineering Faculty, Istanbul University-Cerrahpasa, Turkey, since 2002. Dr. Deniz has twenty-three years of teaching and research experience. She has published more than 60 original scientific research papers indexed in the Science Citation Index, over 120 scientific research communications at national and international conferences, and research articles (indexed in Scopus), including books, book chapters, and patents. She has trained and advised many undergraduate, graduate, and doctoral students.

# Contents

# Preface

Mass spectrometry has emerged as an indispensable analytical platform within the contemporary scientific landscape, distinguished by its exceptional sensitivity, molecular specificity, and capacity for comprehensive structural elucidation. Over recent decades, sustained advancements in instrument design, ionization strategies, separation technologies, and computational methodologies have markedly broadened the analytical capabilities of mass spectrometry. These developments have facilitated its integration into a wide range of disciplines, including, but not limited to, proteomics, metabolomics, pharmaceutical and biomedical research, environmental sciences, petrochemical analysis, and materials characterization.

The present book, *Mass Spectrometry – Applications and Recent Advances*, seeks to provide a rigorous and authoritative examination of the technique's evolving applications and the significant innovations that continue to refine its performance. Particular attention is devoted to recent progress in high-resolution mass analyzers, hybrid and multidimensional platforms, ion mobility spectrometry, imaging mass spectrometry, and advanced quantification strategies. Collectively, these developments underscore the field's dynamic and multidisciplinary nature.

This book is intended for researchers, scholars, and professionals who require a comprehensive, up-to-date synthesis of both fundamental principles and cutting-edge advancements. The insights presented herein will not only inform current scientific practice but also inspire further innovation, thereby contributing to the continued evolution of mass spectrometry as a pivotal tool in analytical science.

**Prof. Dr. Nahide Gülşah Deniz**
Department of Chemistry,
Istanbul University-Cerrahpaşa,
Istanbul, Türkiye

Chapter 1

# Advances in the Analysis of Contaminants, Illicit Drugs, Pharmaceuticals, and Hormones for Wastewater-Based Epidemiology (WBE) Using Automated Extraction and Detection by LC and GC-MS/MS: Evaluation of Method Eco-Friendliness and Practicality

*Dnyaneshwar Shinde, Bhaskar Karubothula, Parth Gupta, Raghu Tadala, Alanood Almansoori, Maryam AlQaydi, Samara Bin Salem, Wael Elamin and Grzegorz Brudecki*

**Abstract**

Wastewater-based epidemiology (WBE) has emerged as a powerful tool for bridging critical data gaps, identifying emerging public health threats, and supporting targeted interventions. This proactive approach for health monitoring ensures early detection and mitigation of risks prior to any escalations. Despite its significance, WBE has traditionally been constrained by obsolete methodologies that require large sample volumes, excessive solvent use, and laborious processes. Existing techniques for detecting chemical contaminants, including semi-volatile and volatile organic compounds (SVOCs and VOCs), illicit drugs (IDs), pharmaceutical and personal care products (PPCPs), and steroids and hormones (SHs), face limitations such as low throughput, restricted applicability to diverse contaminants, and concerns over environmental sustainability. This chapter introduces automated extraction methods to address these challenges, including RTC-PAL-SPME Arrow for SVOCs, Headspace-Trap for VOCs, and the Biomek i7 modular automated workstation with 96-well solid-phase extraction (SPE) plates for IDs, PPCPs, and SHs. Also, this chapter focuses on detection using the latest models of analytical techniques, including APGC-MS/MS, GC-MS, and LC-MS/MS. The proposed advanced analytical approaches were verified at the limits of quantification (LOQ) for representative compounds from each class. Further, their advantages were evaluated for eco-friendliness, using the Green Analytical Procedure Index (GAPI) and Analytical Greenness (AGREE) and practicality was assessed through the Blue Applicability Grade Index (BAGI). Integrating automated sample extraction with the latest models of mass spectrometers coupled to LC/GC sets a new benchmark for WBE, enabling laboratories to adopt more efficient, sustainable, and scalable analytical approaches.

**Keywords:** semi and volatile organic compounds, illicit drugs, pharmaceutical and personnel care products, steroids and hormones, WBE, LC-MS/MS, GC-MS/MS, automated extraction, GAPI, AGREE, BAGI

## 1. Introduction

Wastewater-based epidemiology (WBE) has long served as a tool for monitoring human-derived substances, including pharmaceuticals, illegal drugs, chemicals, and both natural and anthropogenic contaminants. WBE is now widely applied to monitor a broad range of chemical contaminants. Initially focused on detecting pharmaceutical residues in surface and sewage waters, this field has evolved into a globally recognized approach for assessing population-wide exposure and consumption patterns [1]. As its scope continues to expand, there is a growing demand for real-time, high-throughput analysis of urban wastewater. This requires advanced analytical techniques capable of detecting a wide array of compounds, including semi-volatile and volatile organic compounds (SVOCs and VOCs), illicit drugs (IDs), pharmaceuticals and personal care products (PPCPs), as well as steroids and hormones (SHs). However, existing WBE methodologies often face limitations with respect to throughput, analytical suitability for a wide range of chemical classes, and environmental sustainability.

High-throughput analytical methods are essential for real-time analysis and effective WBE implementation. The accurate and timely detection of SVOCs, VOCs, IDs, PPCPs, and SHs in environmental samples presents considerable analytical challenges due to the complexity of matrices, the presence of interferences, and the need for high-sensitivity determinations [2]. Traditionally, in general, all these hazardous chemicals and contaminants in water and other matrices are analyzed using standard methods involving liquid-liquid extraction (LLE) or solid-phase extraction (SPE), followed by detection *via* GC-MS/MS (electron impact (EI) source) or LC-MS/MS (electro spray ionization (ESI) source). These conventional methods typically require large sample and solvent volumes, repeated extraction steps, filtration, cleanup (e.g., GPC, column chromatography, or SPE), solvent evaporation, and sample reconstitution, making them laborious and time-consuming [3]. Also, mass spectrometric analysis proposed in these conventional methods has limitation such as hard ionization (EI source for SVOCs and VOCs, which breaks molecules excessively in the source), extraction and analysis using multiple process (due to pH dependent extraction and separate analysis in negative and positive mode of ionization for IDs and PPCPs), and requirement of derivatization of molecules to make it suitable for detection by LC-MS/MS for SHs [3–5]. Recent advances in extraction and instrumentation raise the following key questions:

1. Can we eliminate the need for repeated extractions of large sample volumes?

2. Can carcinogenic solvents be replaced with greener alternatives?

3. Can an aliquot from settled or centrifuged samples eliminate the need for filtration?

4. Can direct injections without evaporation and reconstitution fit the purpose?

Integrating automation into sample extraction and the use of the latest mass spectrometer models has the potential to overcome these limitations of conventional methods to enhance the efficiency of WBE [6]. Further, the importance of sample preparation in developing greener analytical methods continues to be the focus of many studies [7]. In line with the growing emphasis on green chemistry principles, the high-throughput approaches need to be assessed for method greenness and practicality. Major aspects of the eco-friendly and green sample preparation include the use of eco-friendly solvents, reagents, and materials that are renewable, recyclable, and reusable, minimizing waste generation and energy demand; and enabling high-throughput, miniaturization, procedure simplification, automation, and operator safety [7]. High-throughput analysis must balance speed, ease of operation, safety, cost, and method validation. Additionally, environmental and legal restrictions on hazardous chemicals, especially carcinogenicity, must be considered, as their use is increasingly limited in many countries [8].

Therefore, advancing WBE requires the development of sensitive, robust, and eco-friendly analytical techniques capable of handling high sample throughput, providing reliable data across a wide range of chemical substances, and ensuring data accuracy and consistency. This chapter focuses on advanced, automated analytical methods that are high-throughput, resource-efficient, safe for health and the environment, and maintain accuracy, precision, specificity, and sensitivity as per international regulatory requirements. This chapter explains automated extraction methods, including RTC-PAL-SPME Arrow for SVOCs, Headspace-Trap for VOCs, and the Biomek i7 workstation with 96 well SPE plates for IDs, PPCPs, and SHs. Detection using the latest models of APGC-MS/MS, GC-MS, and LC-MS/MS was assessed for its fitness, as well as the method's environmental sustainability and practicality.

## 2. High-throughput analytical approaches

### 2.1 Automated extraction and analysis of semi-volatile organic compounds (SVOCs) using APGC-MSMS with DH-RTC PAL-SPME Arrow

#### *2.1.1 Introduction*

The analysis of SVOCs in water and other matrices typically follows standard methods [9–21], which involve liquid-liquid extraction (LLE) or solid-phase extraction (SPE), followed by detection using GC-MS (EI source) or LC with specific detectors (ECD, FPD, NPD, and PID). These conventional workflows are labor-intensive and rely on large volumes of carcinogenic solvents such as dichloromethane (DCM), followed by filtration, cleanup (GPC, chromatography), evaporation, and reconstitution [3]. The use of DCM presents environmental and regulatory concerns, with the U.S. Environmental Protection Agency (EPA) finalizing a ban on most uses as of April 30, 2024, due to its hazardous nature [8]. GC-MS/MS with EI ionization is commonly used for SVOC analysis due to its broad applicability. However, the extensive fragmentation caused by EI often results in the loss of molecular ions, reducing method sensitivity and specificity in the case of isotopic or structurally similar compounds. Due to the absence of quasi-molecular ions, method sensitivity and specificity are largely affected [8]. It has been observed in the literature that, in addition to the selectivity advantage arising from the ability to use the quasi-molecular ion, sensitivity is also substantially improved with the use of APGC-MS/MS [22].

The solid-phase micro extraction (SPME) method, developed by Janusz Pawliszyn in 1990, is a solvent-free extraction technique that extracts analytes from a sample with an extraction phase coated on a small diameter fused silica fiber [23]. The phase-coated fiber can be used for Direct Immersion (DI-SPME) in a liquid sample or for Headspace (HS-SPME) sampling where fiber is exposed to the liquid sample or gas phase above the sample in a vial. Over the past few decades, the use of SPME fiber has become one of the most widely used extraction technologies for environmental, food, and clinical analysis. However, the technology remained almost unchanged, with some drawbacks, such as the limited mechanical stability, small phase volumes, and requiring manual interventions for sample extraction and injection into the equipment.

Recently, SPME Arrow, a new patented design for microextraction, has been launched with increased trace-level sensitivity and mechanical durability. With the advanced design of SPME Arrow, it has a larger outside diameter (from 1.1 to 1.5 mm), resulting in large sorption phase surfaces and volumes, and an arrow-shaped tip, which allows penetration of vial and injector septa without damage to SPME phase. With the advancements in instrumentation, SPME Arrow-based analysis evolved with time by combining sample preparation and sample introduction systems on one device. The modular autosampler dual head robotic tool change (DH-RTC) prep and load (PAL) (CTC Analytics AG, Zwingen, Switzerland) equipped with tools for liquid injection, SPME Arrow extraction using a heated agitator and stirrer, a vortex, a conditioning station, and a washing station, has the capacity to automate the sample preparation and sample introduction on APGC-MS/MS (**Figure 1**) without manual intervention.

Based on the principle of adsorption or absorption and desorption, SPME uses a coated fiber or Arrow to concentrate SVOCs from a sample. In SPME, analytes establish equilibria among the sample matrix, the headspace above the sample, and the polymer-coated fused phase. SPME Arrows with different commercial polymer-coated fused phases (PDMS, polyacrylate, carbon WR/PDMS, DVB/PDMS and DVB/carbon WR/PDMS) either in single or in combination of phases are available, which need to be selected based on different selection criteria such as molecular weight of the target analytes, polarity of the target analytes, concentration of the analytes, and the complexity of the matrix. SVOCs include different classes of chemical contaminants and are broadly classified into polycyclic aromatic hydrocarbons (PAHs), organochlorine pesticides (OCPs), polychlorinated biphenyls (PCBs), phthalates, phenols, dioxins and furans (TCDDs and TCDFs), and other semi-volatile organic compounds (SVOCs). For accurate analysis of these compounds, it is essential to pragmatically optimize various parameters (**Table 1**). These include selection of SPME Arrow (phase and thickness), analytes (concentration and compatibility), sample (volume, pH, matrix effect, ionic strength), penetration depths (vial and injector), speed (agitation and stirring), time-temperature (arrow conditioning, sample incubation, extraction and desorption), injection mode (split/splitless), liner (split/splitless/ID-1.7 mm/4 mm), and selection of internal standards.

A fully automated SPME Arrow-based extraction method coupled with atmospheric pressure gas chromatography-tandem mass spectrometry (APGC-MS/MS) using a DH-RTC PAL autosampler was evaluated for its performance, greenness, and practicality. Six representative contaminants from diverse chemical classes were selected to assess the method's suitability. Evaluation parameters included the presence of quasi-molecular ions in the MS spectra, consistency of retention time (RT) and ion ratio, linearity, mean % recovery, and precision at the limits of quantification. The selected compounds reflect both the technical complexity and regulatory requirements of various analytes and their relevance in assessing the operational health of the APGC-MS/MS system. These compounds serve as robust performance indicators for validating the method for compounds across a wide range of chemical properties and regulatory requirements. Selected compounds include:

- *Nitrobenzene:* Early-eluting and low-boiling nitro-aromatic compound commercially used to evaluate the performance of SPME Arrows.
- *Pentachlorophenol:* Phenolic compound well known for its chromatographic challenges and use for calculation of the tailing factor in standard reference methods.
- *Hexachlorobenzene:* Historically used as a reference compound for sensitivity specifications of GC.
- *Gamma-BHC (Lindane):* An organochlorine pesticide representing HCH isomers, selected for its low regulatory limits in different aqueous matrices.
- *PCB 138:* Chosen to represent indicator PCBs, due to its low volatility, congener similarity, and stringent regulatory requirement.
- *Benzo[a]pyrene:* A Group 1 carcinogen and a high-boiling PAH representative of the tailing issues on the instrument, also stringently regulated in water.

### *2.1.2 Analytical method*

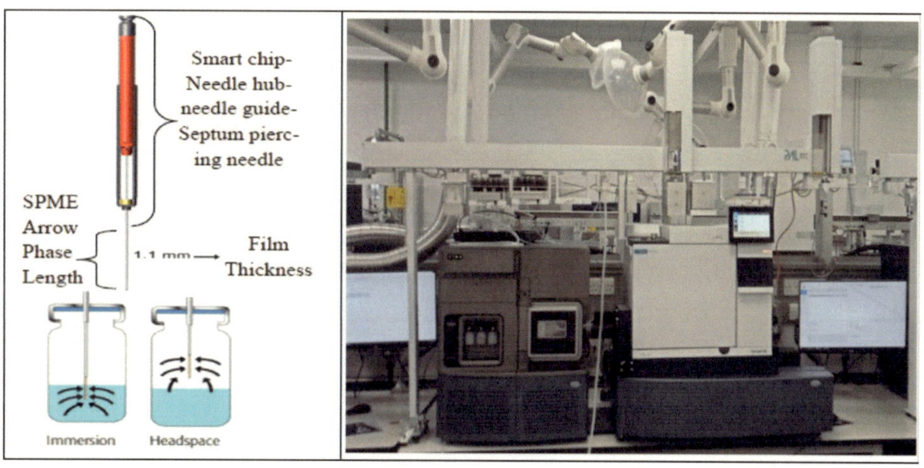


**Figure 1.**
*Configuration of APGC-MS/MS with modular DH-RTC PAL-SPME Arrow.*

| Sr. no. | Automated extraction | | GC method parameters | | MS method parameters | |
|---|---|---|---|---|---|---|
| | DH-RTC PAL with SPME Arrow | | Agilent GC 8890 | | Waters TQ-XS (APCI source) | |
| 1 | Sample | 5 mL (4 g of $Na_2SO_4$) | Inlet type | MMI-SSL | Corona current | 2 µA |
| 2 | SPME Arrow | DVB/PDMS (120 µm × 20 × 1.1 mm) | Liner | Splitless 4.0 mm ID | Auxiliary gas flow | 200 L/hr |
| 3 | Pre-conditioning | 5 min at 260°C under $N_2$ stream | Injection mode | Splitless | Cone gas flow | 260 L/hr |
| 4 | Incubation | 5 min at 80°C, stirring at 500 rpm | Injector temp. | 275°C | Collision gas flow | 0.15 mL/min |
| 5 | Extraction (HS) | 5 min at 80°C, stirring at 250 rpm | Septum purge | 3 mL/min | Source temp. | 150°C |
| 6 | Desorption | GC inlet at 275°C for 9 min | GC column | Select PAH (30 m × 0.25 mm × 0.15 µm) | MS method events | Corona 20 µA (first 3 min), then 2 µA. Cone gas 50 L/hr (first 3 min), then 260 L/hr |
| | | | Carrier gas | He @ 2 mL/min | | |
| | | | Heated transfer line | 330°C | | |
| 7 | Post-conditioning | 5 min at 260°C under $N_2$ stream | GC oven program | 50°C hold for 1 min, 120°C @ 10°C/min, 320°C @ 20°C/min hold 10 min | Acquisition mode | MRM |

**Table 1.**
*Equipment method for automated analysis of SVOCs.*

#### *2.1.3 Results and discussion*

As shown in **Table 2**, quasi-molecular ions were selected as parent ions in the MRM method except in the case of Gamma-BHC (Lindane) because of its labile nature in the source. Ion ratio and RT for selected MRM transitions were observed to be consistent and were meeting the criteria for tolerance of ion ratio and RT, that is, 30% and ±0.1 min, respectively. Extraction method performance for analysis of selected representative compounds was evaluated using isotopically labeled internal standards ($d_5$ nitrobenzene, $^{13}C_6$ pentachlorophenol, $^{13}C_6$ hexachlorobenzene, $d_{10}$ phenanthrene (for lindane), and $d_{12}$ Benzo[a]pyrene). It was observed that the coefficient of determination ($r^2$) ranged from 0.9911 to 0.9974 for hexachlorobenzene and nitrobenzene, respectively. It was observed that the % recovery at respective LOQ ranged from a minimum of 76.39 (for Benzo[a]pyrene) to a maximum of 124.86 (for Lindane). It was observed that precision in terms of % RSD at respective LOQ was ranging from a minimum of 4.89 (for nitrobenzene) to a maximum of 29.27 (for Benzo[a]pyrene).

With the systematic APGC-MS/MS method development for the parameters and use of the Isotopically Labeled Internal Standards (ILIS), the proposed automated analytical method can be considered for qualitative or quantitative analysis of multiclass chemical contaminants in water. In SPME and headspace analysis, the trueness and precision of calibration may depend on the extent to which the analyte has equilibrated with respect to the sample matrix. For SPME Arrow-based automated extraction, as mentioned in SANTE/11312/2021v2 2024 at point number G4, Precision (% RSD) must be demonstrated during verification of method performance [24]. SPME Arrow-based extraction is observed to be suitable for contaminants testing in water, as it meets the requirement of less than 20% RSD for calculated concentration, except for Benzo[a]pyrene (% RSD of 29.27). The optimized method comes with additional benefits of MRM transition with quasi-molecular ion, low toxicity, low cost, suitability for direct injection onto GC, eliminating evaporation, reconstitution steps, and increasing throughput for commercial analysis.

### 2.2 Analysis of volatile organic compounds (VOCs) by headspace (HS)-trap coupled to GC-MS

#### *2.2.1 Introduction*

Volatile organic compounds (VOCs), according to their name, are volatile chemical contaminants and, in almost every case, are analyzed using GC-MS in combination with a sample preparation system. VOCs analysis in water is essential for environmental protection, public health, and regulatory compliance. Conventional reference methods [25–32] with purge-trap and headspace extraction autosampler with GC-MS or flame ionization detector (FID) provide tools for detecting and quantifying VOCs in water. Sample preparation and injection system with GC remains a cornerstone in VOC detection. Some of the commonly used sample preparation and injection systems for VOC analysis include Static headspace (SHS), dynamic headspace (DHS) (also called purge and trap (PT)), thermal desorption, solid-phase microextraction, magnetic solid-phase extraction, and dispersive liquid-liquid microextraction.

| Sr. no. | Name of compounds | SIM method | | Verification of method performance (ng/mL) | | | | |
|---|---|---|---|---|---|---|---|---|
| | | Precursor ion (Q) (Da) | Qualifier ion (q1, q2) | Ion ratio (q1/Q ± Stdev) | RT ± Stdev | $r^2$ | LOQ | % Recovery ± RSD (n = 6) |
| **Representative multiclass volatile organic compounds** | | | | | | | | |
| 1 | Vinyl chloride | 62.0 | 64.0 | 31.752 ± 0.958 | 6.01 ± 0.059 | 0.9804 | 20.0 | 69.44 ± 7.45 |
| 2 | Chloroform | 83.0 | 85.0, 47.0 | 64.829 ± 1.379 | 14.32 ± 0.058 | 0.9985 | 10.0 | 105.90 ± 6.52 |
| 3 | Benzene | 78.0 | 77.0, 51.0 | 23.383 ± 1.638 | 15.32 ± 0.055 | 0.9961 | 10.0 | 122.75 ± 2.65 |
| 4 | Trichloroethene | 95.0 | 130.0 | 75.708 ± 2.553 | 16.50 ± 0.056 | 0.9971 | 10.0 | 122.42 ± 2.42 |
| 5 | Tetrachloroethene | 166.0 | 164.0, 129.0 | 85.963 ± 1.171 | 19.75 ± 0.048 | 0.9938 | 10.0 | 122.40 ± 2.69 |
| 6 | 1,2-Dichlorobenzene | 146.0 | 148.0, 111.0 | 63.850 ± 0.752 | 26.46 ± 0.053 | 0.9996 | 10.0 | 101.60 ± 4.44 |

| Sr. no. | Name of compounds | MRM method | | | | Verification of method performance (ng/mL) | | | | |
|---|---|---|---|---|---|---|---|---|---|---|
| | | Parent ion (Da) | CV (V) | Product ion (q1, q2) | CE (eV) | Ion ratio (q2/q1 ± Stdev) | RT ± Stdev | r2 | LOQ | % Recovery ± RSD (n = 6) |
| **Representative multiclass semi-volatile organic compounds** | | | | | | | | | | |
| 1 | Nitrobenzene | 124.05 | 20 | 77.05, 94.05 | 12, 10 | 0.281 ± 1.91 | 7.51 ± 0.0050 | 0.9974 | 1.0 | 95.83 ± 4.89 |
| 2 | Pentachlorophenol | 265.61 | 20 | 166.70, 201.06 | 25, 20 | 0.488 ± 10.31 | 12.80 ± 0.0048 | 0.9938 | 5.0 | 89.87 ± 14.94 |
| 3 | Hexachlorobenzene | 282.00 | 40 | 248.70, 212.00 | 30, 30 | 0.508 ± 8.57 | 12.79 ± 0.0046 | 0.9911 | 1.0 | 95.14 ± 12.03 |
| 4 | Gamma-BHC | 219.00 | 20 | 182.74, 144.82 | 10, 20 | 0.436 ± 26.11 | 13.43 ± 0.0065 | 0.9954 | 0.5 | 124.86 ± 13.79 |
| 5 | PCB 138 | 361.60 | 40 | 362.10, 360.10 | 10, 5 | 1.337 ± 4.97 | 15.79 ± 0.0075 | 0.9920 | 0.2 | 79.36 ± 13.35 |
| 6 | Benzo[a]pyrene | 253.10 | 40 | 252.10, 250.01 | 35, 52 | 0.218 ± 7.15 | 19.22 ± 0.0065 | 0.9939 | 0.5 | 76.39 ± 29.27 |
| **Representative multiclass illicit drugs** | | | | | | | | | | |
| 1 | Oxymorphone | 302.098 | 42 | 284.000, 227.023 | 23, 28 | 0.640 ± 5.36 | 1.58 ± 0.000 | 0.9934 | 0.330 | 106.77 ± 4.97 |
| 2 | Amphetamine | 136.000 | 25 | 91.000, 119.000 | 19, 15 | 0.105 ± 5.93 | 2.27 ± 0.002 | 0.9990 | 0.330 | 102.17 ± 8.06 |

| Sr. no. | Name of compounds | MRM method | | | | Verification of method performance (ng/mL) | | | | |
|---|---|---|---|---|---|---|---|---|---|---|
| | | Parent ion (Da) | CV (V) | Product ion (q1, q2) | CE (eV) | Ion ratio (q2/q1 ± Stdev) | RT ± Stdev | r2 | LOQ | % Recovery ± RSD (n = 6) |
| 3 | Barbital | 183.000 | 23 | 42.000, 84.938 | 12, 11 | 0.244 ± 2.89 | 2.94 ± 0.005 | 0.9818 | 9.900 | 100.98 ± 5.13 |
| 4 | Cocaine | 304.115 | 37 | 182.026, 81.958 | 17, 29 | 0.313 ± 7.36 | 3.66 ± 0.005 | 0.9958 | 0.033 | 94.95 ± 7.99 |
| 5 | Diazepam | 286.000 | 25 | 194.000, 155.000 | 30, 30 | 0.534 ± 1.84 | 9.14 ± 0.005 | 0.9984 | 0.726 | 98.71 ± 1.10 |
| 6 | THC | 344.900 | 20 | 326.890, 298.300 | 15, 18 | 0.555 ± 7.08 | 10.10 ± 0.003 | 0.9736 | 9.900 | 109.85 ± 7.62 |
| **Representative multiclass pharmaceuticals and personal care products (PPCPs)** | | | | | | | | | | |
| 1 | Acetaminophen | 152.10 | 35 | 110.14, 64.84 | 14, 25 | 0.1474 ± 4.13 | 2.70 ± 0.01 | 0.9982 | 0.05 | 99.33 ± 5.50 |
| 2 | Minocycline | 458.00 | 20 | 441.00, 352.00 | 15, 30 | 0.2533 ± 6.56 | 3.62 ± 0.01 | 0.9892 | 1.00 | 124.90 ± 3.80 |
| 3 | Naproxen | 231.10 | 18 | 184.99, 169.97 | 12, 24 | 0.3163 ± 13.33 | 6.69 ± 0.01 | 0.9956 | 1.00 | 106.82 ± 17.07 |
| **Multiclass and representative steroids and hormones (SH)** | | | | | | | | | | |
| 4 | Equilin | 266.95 | 35 | 142.88, 264.94 | 35, 35 | 0.6667 ± 4.64 | 2.37 ± 0.02 | 0.9882 | 10.00 | 87.49 ± 3.99 |
| 5 | Progesterone | 315.20 | 38 | 97.00, 109.00 | 22, 24 | 0.8871 ± 5.92 | 3.85 ± 0.01 | 0.9927 | 0.20 | 93.92 ± 8.23 |
| 6 | Stigmasterol | 399.26 | 35 | 95.00, 109.04 | 20, 20 | 0.6754 ± 2.34 | 8.45 ± 0.01 | 0.9859 | 20.00 | 82.94 ± 5.87 |

*SIM—single ion monitoring, LOQ—limit of quantification, $r^2$—coefficient of determination, Stdev—standard deviation, RSD—% relative standard deviation, RT—retention time.*

**Table 2.**
*Evaluation of method performance for automated analysis of VOCs, SVOCs, IDs, PPCPs, and SHs.*

Analyzing volatile organic compounds (VOCs) presents several common challenges that can impact data quality and reliability. These include low sensitivity and poor reproducibility, sample contamination, and carryover between runs. Blockages in the sample pathway, often caused by solid particles, can disrupt analysis, while excessive purging may lead to the loss of highly polar or highly volatile VOCs if not properly optimized and controlled. Foaming during sample introduction further complicates analysis, particularly with wastewater matrices. Additionally, thermally labile VOCs often require larger sample volumes due to their lower sensitivity, and high salinity or organic content in wastewater, which can lead to matrix interferences that must be carefully managed to ensure accurate results. Furthermore, methods using flame ionization detectors (FID) are generally unsuitable for trace-level VOC analysis, as they lack the sensitivity and selectivity required for detecting low concentrations of analytes in complex matrices.

The analysis of volatile organic compounds (VOCs) is often criticized for its limited flexibility in reanalysis. This issue is particularly pronounced when working with small sample volumes or when a repeat analysis is needed, as unstable VOCs may degrade or volatilize during the time taken for the initial examination. The selection of a sample preparation system plays an important role in VOC analysis, and it is a complex decision shaped by several factors, including the sample matrix, target analytes, and specific analytical objectives. While various systems offer distinct advantages, a key limitation persists due to the historical "one-shot" injection approach in VOC testing. Despite their strengths, this limitation emphasizes the importance of choosing a sample preparation technique that balances analytical performance and practicality.

The automated Markes Centri® HS-trap (Markes International Ltd., Bridgend, UK) methodology has the ability of sample enrichment for increasing method sensitivity and reproducibility, along with the capability for sample storage for reanalysis (**Figure 2**). An innovative methodology of HS-trap (by sample re-collection in tubes) allows repeat analysis, which provides more flexibility for VOC testing. The primary benefits of sample re-collection include the ability to archive samples, improve sample characterization, and obtain additional data using thermal desorption (TD) tubes with various adsorbents. Moreover, it extends the dynamic range by employing different split conditions for the original and re-collected samples, which is particularly advantageous for unexpectedly high concentrations that may overwhelm the detector. Additionally, it provides flexibility during the optimization and validation stages of analytical methods.

The sample preparation using the HS-trap system principally involves major steps such as equilibration of samples, loading of the sample on the trap, adsorption/absorption of sample components on the trap, repeating the steps for enrichment, if required, and trap desorption onto to GC column. Specially designed focusing traps equipped with selective stationary phases are used for the efficient adsorption or absorption of VOCs. These traps often contain up to four types of selective sorbent beds, optimized to capture a broad range of VOCs based on their physicochemical properties.

During equilibration, the headspace (HS) vial containing the sample is placed in a heated agitator and maintained at a fixed temperature, agitation speed, and duration to allow equilibrium to be established between the liquid or solid sample and the vapors in the headspace. After the equilibration step, a defined sample volume (gas phase from headspace of the vial) can be injected into the focusing trap

(**Table 3**). For increasing the sensitivity of analysis, focusing traps can be enriched with repeated injection of sample volume after another cycle of equilibration of the respective sample vial. After sample loading, the trap undergoes rapid heating to a designated desorption temperature, enabling the swift and efficient release of the trapped analytes for subsequent analysis. The analytes adsorbed onto the trap are desorbed by rapidly heating the trap to high temperatures while passing the carrier gas through the trap to the column. Six representative VOCs were selected to assess the performance of this method. Method performance parameters evaluated were the presence of SIM ions in the MS spectra, consistency of retention time (RT) and ion ratio, linearity, mean % recovery, and precision at the limits of quantification.

Selected compounds include:

- *Vinyl chloride:* Group A confirmed carcinogen, part of Directive (EU) 2020/2184, suitable for evaluating chromatography of early-eluting VOCs.

- *Chloroform:* Group 2B possibly carcinogenic, part of Directive (EU) 2020/2184, important due to frequent detection as disinfection byproduct.

- *Benzene:* Group A confirmed carcinogen, part of Directive (EU) 2020/2184, and low regulatory limits reflect the high toxicity.

- *Tetrachloroethene:* Group 2A possibly carcinogenic, part of Directive (EU) 2020/2184, and critically important due to widespread industrial use.

- *Trichloroethene:* Group A confirmed carcinogen, part of Directive (EU) 2020/2184 [33] and critically important due to its frequent detection.

- *1,2-Dichlorobenzene:* Suitable for evaluating chromatography of late-eluting VOCs, and is important as it is regulated for industrial discharge permits.

*2.2.2 Analytical method*

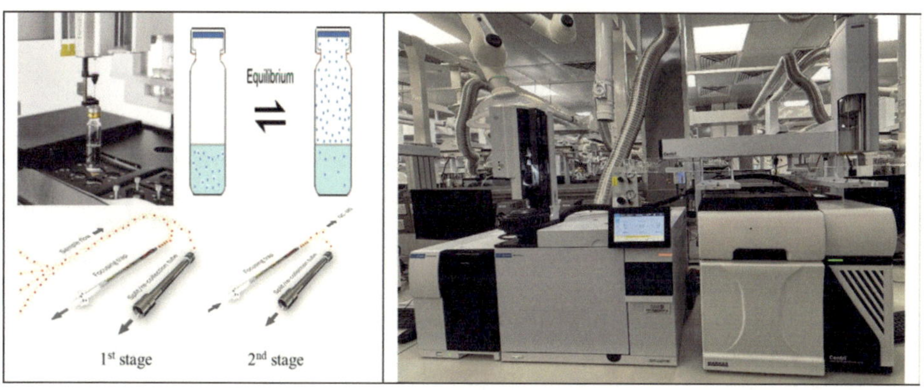


**Figure 2.**
*GC-MS with HS-trap.*

| Sr. no. | Automated extraction | | GC-MS method parameters | |
|---|---|---|---|---|
| | **HS-trap parameters** | | **Agilent 8890 GC with 5977B MSD** | |
| 1 | HS syringe temperature | 90°C | Column | VF624ms, 60 m × 250 μm × 1.4 μm |
| 2 | Sample volume | 10 mL | GC oven program | 35°C hold for 6 min, 8°C per min to 240°C hold for 2 min |
| | HS sample volume | 2.5 mL | | |
| 3 | Incubation temperature | 80°C | Carrier gas | He (Helium) |
| 4 | Incubation time | 10 min | Column flow | 1.2 mL/min |
| 5 | Injector temperature | 200°C | Inlet type | SSL |
| 6 | Injection time | 2 min | Inlet mode | Split less |
| 7 | Focusing trap | TO-15/TO-17 Air toxics trap | Liner | Splitless 4.0 mm id |
| 8 | Trap purge time | 1 min | Ionization | Electron impact |
| 9 | Purge flow | 50 mL/min | MS source | 70 eV |
| 10 | Trap low | 15°C | Source temperature | 230°C |
| 11 | Trap heating rate | Max (~100°C/sec) | Quadruple temperature | 150°C |
| 12 | Trap high | 280°C | Transfer line temperature | 220°C |
| 13 | Trap desorb time | 3 min | Solvent delay | 0.5 min |
| 14 | Split flow | 5 mL/min | Acquisition mode | SIM |
| 15 | Total number of injections | 2 | Mass range | m/z 45–300 |
| 16 | Enrichment delay (min) | 6.0 | Tune type | E-tune |

**Table 3.**
*Equipment method for automated analysis of VOCs.*

### *2.2.3 Results and discussion*

As shown in **Table 2**, a quasi-molecular ion was selected as a precursor ion in the SIM method except in the case of trichloroethene and chloroform due to their labile nature in the electron impact source, due to dissociation of chlorine groups. In the case of trichloroethene, quasi-molecular ion was used as a qualifier, as it was the second intense ion in the EI source mass spectra. The ion ratio and RT for selected SIM ions were observed to be consistent and met the acceptable tolerance of 30% and ±0.1 min, respectively. Extraction method performance for analysis of selected representative compounds was evaluated using isotopically labeled internal standards ($^{13}$C Chloroform, Benzene-$d_6$, and 1,2-dichlorobenzene-$d_4$). It was observed that the coefficient of determination ($r^2$) ranged from a minimum of

0.9804 to 0.9996 for vinyl chloride and 1,2-dichlorobenzene, respectively. It was observed that the mean % recovery at respective LOQ ranged from a minimum of 69.44 to a maximum of 122.75 for vinyl chloride and benzene, respectively. It was observed that precision in terms of % RSD at respective LOQ was ranging from a minimum of 2.42 to a maximum of 7.45 for trichloroethene and vinyl chloride, respectively.

With the systematic GC-MS method development, evaluation of parameters affecting HS-trap extraction, and use of the Isotopically Labeled Internal Standards (ILIS), the proposed automated analytical method can be considered for quantitative analysis of multiclass VOCs in water. In the case of headspace analysis, the trueness and precision of calibration may depend on the extent to which the analyte has equilibrated with respect to the sample matrix. For HS-based automated extraction as mentioned in SANTE/11312/2021v2 2024 at point number G4, Precision (% RSD) must be demonstrated during verification of method performance [24]. HS-trap-based extraction is observed to be suitable for VOCs testing in water as it meets the requirement of less than 20% RSD. This multidimensional and modular automated extraction platform overcomes the biggest challenge in the analysis of VOCs that it must be done in "one-shot" (by sample re-collection in tubes, which allows repeat analysis).

## 2.3 Analysis of illicit drugs (IDs) using modular automated sample preparation workstation by LC-MS/MS

### *2.3.1 Introduction*

The estimation of community drug use through wastewater analysis demands high-throughput, accurate, and sensitive quantification methods. A significant challenge in analyzing illicit drugs for subsequent calculations of drug loads in wastewater and drug consumption is the low concentration of these drugs in composite samples collected from various manholes and points within the sewer system. This low concentration necessitates analysis of a large number of samples to avoid sampling artifacts due to variations of wastewater flows and drug loads, which could otherwise lead to over- or underestimation of drugs. Furthermore, the presence of many other substances in the sample matrix hampers the accuracy of the results.

Several LC-MS/MS-based analytical methods for detecting illicit drugs have been reported, with varying degrees of limitations. These methods typically need large sample volumes (50–1000 mL) and are often tested on a few samples and analytes [34–39]. Manual sample preparation and multiple steps hinder high-throughput real-time analysis. Recent methodologies often analyze up to 50 samples using ultrapure water for linearity and quality controls to measure illicit drug concentrations in wastewater. This approach can underestimate the effect of matrix components on result accuracy [40–44]. Hence, detection methodologies used should be specific, efficient, and accurate.

This chapter presents a study of a fully automated robotic system for the extraction and analysis of illicit drugs (IDs) across different classes from wastewater. Sample preparation was automated using a Biomek i7 Workstation, equipped with an automated 8-channel pipetting robotic arm, positive pressure SPE extractor, and multidrop combo. The sample preparation workstation automates tasks such as conditioning of SPE cartridges, sample loading, cleanup, elution, and sealing of 96-well SPE plates. Sample extraction starts with placing SPE plates, solvent reservoirs, and solvents for conditioning, equilibration, washing, and elution on the workstation deck as instructed in the software method (**Figure 3**). The user then selects the tubes containing the samples and controls in the software to provide the pipetting and dispensing locations to the robotics. After the above tasks, the preprogrammed MCX SPE method is then selected to run the complete extraction protocol. All the reagents were dispensed into plates using the 8-channel pipetting tools and by a multidrop combi dispenser by using diverting valve selection through software.

Pre-centrifuged 10 mL wastewater samples were transferred from the sample bottle to a 15 mL polypropylene centrifuge tube using the Hamilton robotics system. To these samples, 50 µL of a deuterated internal standard mix and 500 µL of 2% formic acid in ASTM Type I water were added manually. Then, samples, calibration controls (CCs), and quality controls (QCs) were mixed thoroughly using a benchmark multi-tube vortexer. The samples ready for the automated solid-phase extraction need to be loaded onto the Biomek i7 workstation for solid-phase extraction (SPE). **Table 4** outlines the steps for the automated extraction of illicit drugs in wastewater. The optimized MS method was tested for both negative (barbiturates) and positively ionized molecules by polarity switching to eliminate separate analysis of negatively charged molecules. This fully automated modular sample preparation workstation (Biomek i7) using MCX 96-well plates 60 µm (60 mg) SPE cartridges with abilities of very high throughput and without evaporative preconcentration of sample is evaluated for its performance. Evaluation parameters included the presence of SIM ions in the MS spectra, consistency of retention time (RT) and ion ratio, linearity, mean % recovery, and precision at the limits of quantification. The most relevant and challenging six representative IDs from different classes were selected to assess the method's suitability.

Selected compounds include:

- *Oxymorphone:* It is an early-eluting compound from the opioid class, and it is challenging due to coelution and matrix effect.
- *Amphetamine:* This class of synthetic drugs production and presence is rapidly growing and there is a need for high-throughput analytical method.
- *Barbital:* This compound is from the barbiturates class and requires a negative mode for its ionization in the source.
- *Diazepam:* It is a prescribed drug from the benzodiazepine class, and it undergoes biotransformation and degradation to its metabolites.
- *Cocaine:* It is well known, most frequently monitored, and is one of the compounds pioneering the use of WBE in the early 2000s.
- *11 Nor 9 Carboxy Delta 9 tetrahydrocannabinol (THC):* It is well known, frequently monitored, and represents late-eluting IDs.

*DOI: http://dx.doi.org/10.5772/intechopen.1011550*

### *2.3.2 Analytical method*

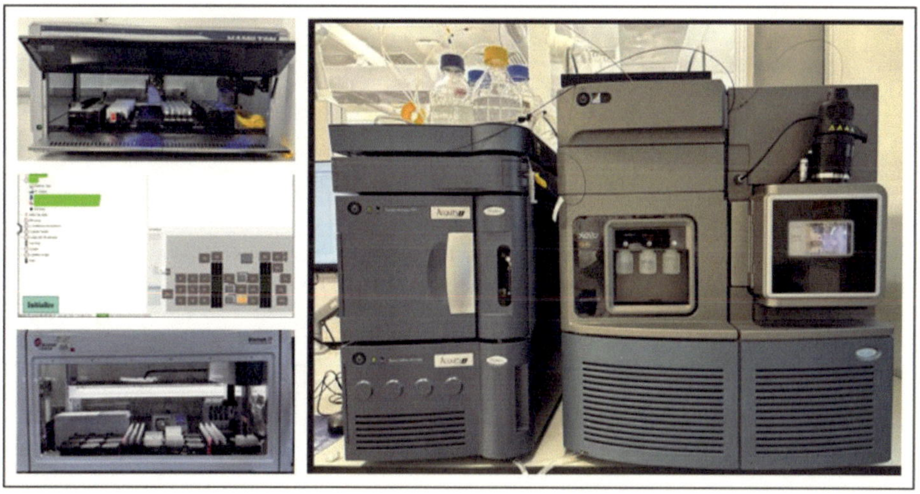

**Figure 3.**
*LC-MS/MS with Biomek i7 Workstation for automated sample preparation and Hamilton robotics system for sample aliquoting.*

| | Steps | Sample preparation task |
|---|---|---|
| Procedure for automated extraction of IDs in wastewater using Biomek i7 | Sample preparation | Unknown wastewater samples and Spiked (QC) wastewater samples with ISTD and extraction buffer were loaded in the 15 mL conical tube racks (samples 10.0 mL) and precleared by centrifugation at 4200 rpm in a model (Eppendorf). |
| | Conditioning | Condition the MCX 96-well plate cartridge with 1.0 mL 100% MeOH |
| | Equilibration | Equilibrate the cartridge with 1.0 mL 2% formic acid in ASTM Type 1 water. |
| | Sample loading | Load 1.5 mL of samples and quality controls three times on the respective MCX 96-well plate cartridge. |
| | Washing | Wash all the cartridges with 1.0 mL of 2% formic acid in ASTM Type 1 water. Dry the cartridge at 1000 mbar for up to 10 min. |
| | Elution | Elute with 500 μL of 5% ammonium hydroxide in MeOH: ACN (50:50 v/v). |
| | Dilution and well plates sealing | Add 200 μL of mobile phase buffer to the elution plate, vortex, and seal the MCX 96-well plate cartridge with a4S 4titude automated role heat sealer. |
| | LC-MS/MS detection | Load sealed MCX 96-well plate cartridge LC-MS/MS autosampler and inject 4 μL of sample volume. |
| Equipment details | Ion source | Z Spray XEVO ion source |
| | Pump | Acquity UPLC I-Class plus |
| | Auto sampler | FTN sample manager |
| | Column oven | Acquity UPLC column heater |
| | LC column | Acquity Premier UPLC BEH C18 column, 1.7 μm, 2.1 × 100 mm |

| | Steps | Sample preparation task | | |
|---|---|---|---|---|
| LC parameters | Mobile Phase A | 0.50 mM ammonium acetate with 0.025% acetic acid in water | | |
| | Mobile Phase B | MeOH: ACN (70/30, v/v) with 0.025% acetic acid | | |
| | Sample purge | ACN: Water (50/50, v/v) | | |
| | Sample wash | ACN: Water (50/50, v/v) | | |
| | Seal wash | MeOH: Water (90/10, v/v) | | |
| | Flow rate | 0.350 mL/min | | |
| | Column oven | 55 ± 5°C | | |
| | Sample manger | 8 ± 5°C | | |
| | Injection | 4.0 μL | | |
| LC gradient | **Flow** | **Time** | **Pump A%** | **Pump B%** |
| | 0.35 | 0 | 98 | 2 |
| | 0.35 | 0.5 | 98 | 2 |
| | 0.35 | 0.6 | 85 | 15 |
| | 0.35 | 8 | 45 | 55 |
| | 0.35 | 9.5 | 2 | 98 |
| | 0.35 | 10.5 | 2 | 98 |
| | 0.35 | 10.6 | 98 | 2 |
| | 0.35 | 12 | 98 | 2 |
| MS parameters | | Mode and polarity | | ESI +/– |
| | | Scan type | | (MRM) |
| | | Source temperature (°C) | | 150 |
| | | Desolvation gas temp. (°C) | | 600 |
| | | Desolvation gas flow (L/hr) | | 1100 |
| | | Cone gas flow (L/hr) | | 150 |
| | | Capillary voltage (kV) | | 1.6 |
| | | Nebulizer gas flow (bar) | | 7 |

**Table 4.**
*Method for automated analysis of illicit drugs using Biomek i7 and LC-MS/MS.*

### *2.3.3 Results and discussion*

As shown in **Table 2**, quasi-molecular ions were selected as parent ions in MRM method for all selected IDs. All six targeted representative IDs (early and last eluting from different classes) with their internal standards (ISTDs) were eluted within 10.1 min of runtime, the remaining 1.40 min of runtime were used for proper column equilibration to make it ready for the next injection. The ion ratio and RT for selected MRM transitions were observed to be consistent and met the criteria for tolerance of ion ratio and RT, that is, 30% and ± 0.2 min, respectively. The performance of the extraction method for the analysis of selected representative compounds was evaluated. The coefficient of determination ($r^2$) was observed to range from 0.9736 for THC to 0.9990 for amphetamine. The % recovery at the respective limit of quantitation (LOQ) was observed to range from a minimum of 94.95% for cocaine to a maximum of 109.85% for THC. The precision, expressed as % RSD at respective LOQ, ranged from 1.10 for diazepam to 8.06 for amphetamine.

With systematic LC-MS/MS method development and evaluation of parameters affecting MCX 96-well plate solid-phase extraction (SPE) and use of Isotopically Labeled Internal Standards (ILIS) (oxymorphone $d_3$, amphetamine $d_6$, cocaine $d_2$, and THC $d_3$), the proposed automated analytical method can be used for quantitative analysis of multiclass IDs. The method demonstrated good sensitivity with LOQ of 0.033 (cocaine), 0.330 (amphetamine and oxymorphone), 0.726 (diazepam), 9.9 (barbital and THC) ng/mL with a small sample volume of 4.5 mL (**Table 2**). Further, with the switching of the polarity, both negatively and positively ionized molecules were analyzed in a single run.

## 2.4 Analysis of pharmaceuticals and personal care products (PPCPs) including steroids and hormones (SHs) using modular automated sample preparation workstations by LC-MS/MS

### *2.4.1 Introduction*

Pharmaceuticals and personal care products (PPCPs), including steroids and hormones (SHs), are a group of potential water pollutants that are categorized as contaminants of emerging concern (CECs). These compounds, under the broad umbrella of PPCPs, serve a wide array of purposes, such as antibacterials, antiseptics, preservatives, therapeutics, and disinfectants, and are found in countless healthcare products. In recent decades, the production and consumption of pharmaceutical products have rapidly increased, with approximately 3000 compounds developed, and the annual production exceeds hundreds of tons [45]. Despite their benefits, concerns remain about the continuous discharge of these substances into the environment through multiple pathways, like pharmaceutical industries, hospitals, improper disposal, and inadequate wastewater treatment by wastewater treatment plants (WWTPs) [46]. These compounds can find their way into the sea or surface water through stormwater runoff, household waste, sewage, bathing/showering, and direct disposal. Contaminated sea or surface water used as a source for drinking water, upon treatment, can cause concerns about the quality of drinking water.

Studies have reported the presence of CECs in water; the occurrence of CECs is of concern because they remain active even at trace concentrations. Several studies have suggested that long-term exposure to CECs could cause allergies, antibiotic resistance, and effects on the endocrine system. Several studies have identified a growing resistance pattern to antibiotics due to the increased use of existing antibiotics and posing the biggest threat to humans in the form of antimicrobial resistance (AMR). According to the World Health Organization (WHO), overuse and misuse of antimicrobials are the main drivers in the development of drug-resistant pathogens. Antimicrobial resistance (AMR) is one of the most pressing global health threats of the twenty-first century. A recent study projects that between 2025 and 2050, AMR can cause 39.1 million deaths [47]. These figures are alarming, and WBE is a great tool that can provide evidence for existing data gaps, helping to expand understanding of this threat and support targeted interventions.

There is a need to monitor contamination of these CECs in water, and currently some PPCPs are listed on U.S. EPA Contaminants Candidate List (CCL) 3 and CCL 4, indicating their presence in public water systems and potential need for regulation under the Safe Drinking Water Act (SDWA) [48, 49]. In 2007, the Environmental Protection Agency (EPA) published reference method: Method 1694: "Pharmaceuticals and Personal Care Products in Water, Soil, Sediment, and Biosolids by HPLC/MS/MS." This method divides PPCPs into acid and base fractions and uses two extraction protocols. Further, target analytes in this method are divided into four groups. Groups 1, 2, and 3 are extracted under acidic (pH 2) conditions. Groups 1 and 2 are run in the positive electrospray ionization ($ESI^+$) mode, and Group 3 is run in the negative electrospray ionization ($ESI^-$) mode. Group 4 is extracted under basic (pH 10) conditions and is run in the $ESI^+$ mode. Further, a very high sample volume of 1000 mL must be passed through an HLB 20 cc 1 g cartridge and concentrated to 3 mL [4].

Method 1698: "Steroids and Hormones in Water, Soil, Sediment, and Biosolids by HRGC/HRMS." In this method, a very high sample volume of 1000 mL was extracted using methylene chloride (carcinogenic solvent) in a separatory funnel. Further, the samples are subjected to cleanup as required using a layered alumina/Florisil column. Following cleanup, the sample extract of about 320 mL of methylene chloride is concentrated to approximately 0.1 mL, which is highly unsafe. The solvent is exchanged to pyridine, and the steroids/hormones are derivatized to their trimethylsilyl-ethers using N, O-Bis(trimethylsilyl) trifluoroacetamide with trimethylchlorosilane (BSTFA: TMCS) [5].

These methods recommend large sample volumes and hazardous chemicals to meet the required sensitivity. Additionally, separate injections were required for negatively and positively ionized molecules because mass spectrometers in the early 2000s could not efficiently analyze compounds in polarity switching mode during a single analytical run. The extraction of high sample volume and evaporative concentration of large extraction solvent is laborious, time-consuming, error-prone, and unsafe for humans and the environment. Further, the step derivatization using BSTFA: TMCS for SHs requires extra time, increases uncertainty, and raises safety issues.

The automated modular sample preparation workstation (Biomek i7) utilizing 96-well plate SPE cartridges and LC-MS/MS with ESI and APCI source of ionization (as shown in **Figure 3**) offers high-throughput capabilities without the need for evaporative preconcentration, derivatization, or high sample volume. Method performance is assessed accordingly for the most relevant and challenging six representative PPCPs and SHs from different classes. Evaluation parameters included the presence of molecular ions in the MS spectra, consistency of retention time (RT) and ion ratio, linearity, mean % recovery, and precision at the limits of quantification. These selected compounds include:

#### *2.4.1.1 Pharmaceutical and personal care products*

*Acetaminophen*: It is a widely used pharmaceutical compound and represents chromatographically early-eluting compounds.

*Minocycline*: It is an unstable pharmaceutical compound, prone to degradation due to changes in temperature and pH, and also gets affected heavily due to the matrix effect.

*Naproxen*: It is a commonly used pharmaceutical compound and represents chromatographically late-eluting compounds.

#### *2.4.1.2 Steroids and hormones*

*Equilin*: It is a naturally occurring estrogenic hormone and represents well chromatographically mid-eluting and negatively ionized compound.

*Progesterone*: It is a natural steroid hormone and represents well chromatographically mid-eluting compounds.

*Stigmastanol*: It is a plant sterol and represents well chromatographically late-eluting compounds.

### *2.4.2 Analytical method*

See **Table 5**.

*DOI: http://dx.doi.org/10.5772/intechopen.1011550*

| | Steps | Sample preparation task |
|---|---|---|
| Procedure for automated extraction of IDs in wastewater using Biomek i7 | Sample preparation | Unknown wastewater samples and spiked (QC) wastewater samples with ISTD and extraction buffer (0.400 mL of 5% ammonia in water only in case of PPCPs) were loaded in the 15 mL conical tube racks (samples 10.0 mL) and precleared by centrifugation at 4500 (Eppendorf). |
| | Conditioning | **PPCPs:** Condition the MAX 96-well plate cartridge with 1.0 mL 100% MeOH. |
| | | **SHs:** Condition the HLB 96-well plate cartridge with 1.0 mL 100% MeOH. |
| | Equilibration | Equilibrate the cartridge with 1.0 mL of 100% ASTM Type I water. |
| | Sample loading | Load 1.5 mL of samples and quality controls three times (total sample volume 4.5 mL) on the respective 96-well plate cartridge. |
| | Washing | **PPCPs**: Wash all the cartridges with 1.0 mL of 5% ammonia in ASTM Type I water. Dry the cartridge at 1000 mbar for up to 10 min. |
| | | **SHs:** Wash all the cartridges with 1.0 mL of 10% methanol in ASTM Type I water. Dry the cartridge at 1000 mbar for up to 10 min. |
| | Elution | **PPCPs:** Elute cartridge with 0.5 mL of 2% formic acid in methanol: acetonitrile (50:50, v/v). |
| | | **SHs:** Elute the cartridge two times with 0.5 mL of ethyl acetate: n-hexane (70:30 v/v) into the elution plate. |
| | Dilution and well plates sealing | **PPCPs:** Dilute the final elute sample with 0.750 mL of water and seal the cartridge plate by a4S 4titude automated role heat sealer and vortex mixture at lower rpm.<br>**SHs:** After drying (at 60°C temperature under stream of nitrogen at gas flow of 60 L/m, for approx. 30 min) completely, reconstitute the well plate with 0.200 mL of methanol: acetonitrile (70:30) v/v and 0.400 mL of ASTM Type I water and seal the cartridge plate by a4S 4titude automated role heat sealer and vortex mixture at lower rpm. |
| | LC-MS/MS detection | **PPCPs:** Load sealed 96-well plate cartridge LC-MS/MS autosampler and inject 4 μL of sample<br>**SHs:** Load sealed 96-well plate cartridge LC-MS/MS autosampler and inject 10 μL of sample |
| Equipment details | Ion source | Z Spray XEVO ion source |
| | Pump | Acquity UPLC I-Class plus |
| | Auto sampler | FTN sample manager |
| | Column oven | Acquity UPLC column heater |
| | LC column | **PPCPs:** ACQUITY Premier UPLC BEH C18 column, 1.7 μm, 2.1 mm × 100 mm |
| | | **SHs:** X Bridge premier BEH C18 column, 2.5 μm, 2.1 mm × 100 mm |

| | Steps | Sample preparation task | |
|---|---|---|---|
| LC parameters | Mobile Phase A | **PPCPs** | **SHs** |
| | | 0.1% acetic acid in water | 0.2 mM ammonium fluoride in water |
| | Mobile Phase B | 0.1% acetic acid in MeOH: ACN (50/50, v/v) | 100% methanol |
| | Sample purge | ACN: MeOH: IPA: Water: (1:1:1:1 with 0.5% acetic acid v/v/v) | ACN: MeOH: IPA: Water: (1:1:1:1 with 0.1% formic acid v/v/v) |
| | Sample wash | MeOH: ACN: Water: IPA (1:1:1:1, v/v) with 0.5% acetic acid | ACN: MeOH: IPA: (20:40:40 with 0.1% formic acid v/v/v) |
| | Seal wash | MeOH: Water (10/90, v/v) | MeOH: Water (90/10, v/v) |
| | Flow rate | 0.35 mL/min | 0.40 mL/min |
| | Column oven | 55 ± 5°C | 65 ± 5°C |
| | Sample manger | 10 ± 3°C | 8 ± 5°C |
| | Injection | 4.0 μL | 10.0 μL |

| **LC gradient** | **PPCPs** | | | | **SHs** | | | |
|---|---|---|---|---|---|---|---|---|
| | Flow | Time (min) | Pump A% | Pump B% | Flow | Time | Pump A% | Pump B% |
| | 0.350 | Initial | 100.0 | 0.0 | 0.4 | Initial | 50 | 50 |
| | 0.350 | 0.60 | 100.0 | 0.0 | 0.4 | 0.10 | 50 | 50 |
| | 0.350 | 2.00 | 80.0 | 20.0 | 0.4 | 3.0 | 35 | 65 |
| | 0.350 | 5.00 | 50.0 | 50.0 | 0.4 | 3.10 | 12 | 88 |
| | 0.350 | 5.50 | 50.0 | 50.0 | 0.4 | 8.5 | 0 | 100 |
| | 0.350 | 5.60 | 30.0 | 70.0 | 0.4 | 9.0 | 0 | 100 |
| | 0.350 | 8.00 | 10.0 | 90.0 | 0.4 | 10.5 | 0 | 100 |
| | 0.350 | 9.00 | 10.0 | 90.0 | 0.4 | 11 | 50 | 50 |
| | 0.350 | 9.10 | 2.0 | 98.0 | 0.4 | 13.0 | 50 | 50 |
| | 0.350 | 10.00 | 2.0 | 98.0 | NA | | | |
| | 0.350 | 10.10 | 100.0 | 0.0 | | | | |
| | 0.350 | 12.00 | 100.0 | 0.0 | | | | |

DOI: http://dx.doi.org/10.5772/intechopen.1011550

| Steps | | Sample preparation task | |
|---|---|---|---|
| MS parameters | Mode and polarity | PPCPs | SHs |
| | | ESI +/− | APCI +/− |
| | Scan type | (MRM) | (MRM) |
| | Source temperature (°C) | 150 | 500 |
| | Desolvation gas temp. (°C) | 600 | 500 |
| | Desolvation gas flow (L/hr) | 1100 | 1000 |
| | Cone gas flow (L/hr) | 150 | 150 |
| | Capillary voltage (kV) | 1.2 | 3 |
| | Nebulizer gas flow (bar) | 7 | 7 |

**Table 5.**
*Method for automated analysis of PPCPs and SDs using Biomek i7 and LC-MS/MS.*

*2.4.3 Results and discussion*

As shown in **Table 2**, quasi-molecular ions were selected as parent ions in the MRM method for all selected PPCPs and SHs. The ion ratio and RT for selected MRM transitions were observed to be consistent and met the criteria for tolerance, that is, 30% and ± 0.2 min, respectively. It was observed that coefficient determination ($r^2$) ranged from a minimum of 0.9859 (Stigmastanol) to a maximum of 0.9982 (Acetaminophen). It was observed that the % recovery at respective LOQs ranged from a minimum of 82.94 (Stigmastanol) to a maximum of 124.90 (Minocycline). It was observed that precision in terms of % RSD at respective LOQ was ranging from a minimum of 3.80 (Minocycline) to a maximum of 17.07 (Naproxen).

With systematic LC-MS/MS method development and evaluation of parameters affecting solid-phase extraction (SPE) and use of Isotopically Labeled Internal Standards (ILIS), the proposed automated analytical method can be used for quantitative analysis of multiclass PPCPs and SHs. The method demonstrated good sensitivity with LOQ of 0.05 (Acetaminophen), 0.20 (Progesterone), 1.00 (Minocycline and Naproxen), 10.00 (Progesterone), 20.00 (Stigmastanol) µg/L with a small sample volume of 4.5 mL. This modular automated extraction protocol achieves the required sensitivity without the need for evaporative concentration for PPCPs and derivatization for SHs.

## 3. Evaluation of method greenness and practicality

### 3.1 Introduction

Traditional methods, including liquid-liquid extraction (LLE) and solid-phase extraction (SPE), typically involve manual, multi-step processes that require substantial amounts of solvents and energy. These methods also produce considerable waste and pose potential health risks to operators. Consequently, sample preparation is frequently regarded as the main contributor to the environmental impact of these techniques [50]. The very first suggestion for applying Green Sample Preparation (GSP) was to use direct analytical techniques. However, this is not feasible, especially when sample conversion into a form suitable for analysis is required [51].

To address this gap, the GSP concept was introduced, providing guidelines for sample preparation that reduce hazardous inputs, minimize waste, and maximize efficiency while preserving analytical performance [51]. Within the framework of GSP, the adoption of microextraction techniques, solvent-free procedures, and automated workflows has demonstrated the feasibility of achieving high analytical performance while significantly reducing environmental impacts. To evaluate and harmonize the compliance of analytical methods with green criteria, several metrics of varying comprehensiveness have been proposed. The proposed methods were evaluated for their greenness using two tools: Green Analytical Procedure Index (ComplexMoGAPI) and Analytical Greenness (AGREE), and their practicability for high-throughput analysis was assessed using Blue Applicability Grade Index (BAGI).

AGREEprep was the first green metric tool designed to evaluate the environmental impact of sample preparation methods [52]. Built upon the 10 principles of GSP, AGREEprep assesses methods across 12 criteria. In 2021, an enhanced

*DOI: http://dx.doi.org/10.5772/intechopen.1011550*

ComplexMoGAPI framework was reported by adding a hexagonal segment to assess activities beyond sample preparation and final analysis, considering parameters like recovery yields, processing conditions, green economy principles, occupational hazards, and E-factor for relevant methods [53]. BAGI was introduced by Manousi et al. in 2023 as a novel metric for assessing the practicality of analytical methods [54]. Unlike traditional green metrics that focus on environmental impact, BAGI centers on 10 key attributes crucial for method applicability [54].

### 3.2 Results for evaluation of greenness and practicality

Modified version of a software, ComplexMoGAPI, was used to simplify the assessment with a scoring system wherein a score of 100 represents the method is 100% green, excellent if greater than 75, acceptable if greater than 50, and not acceptable

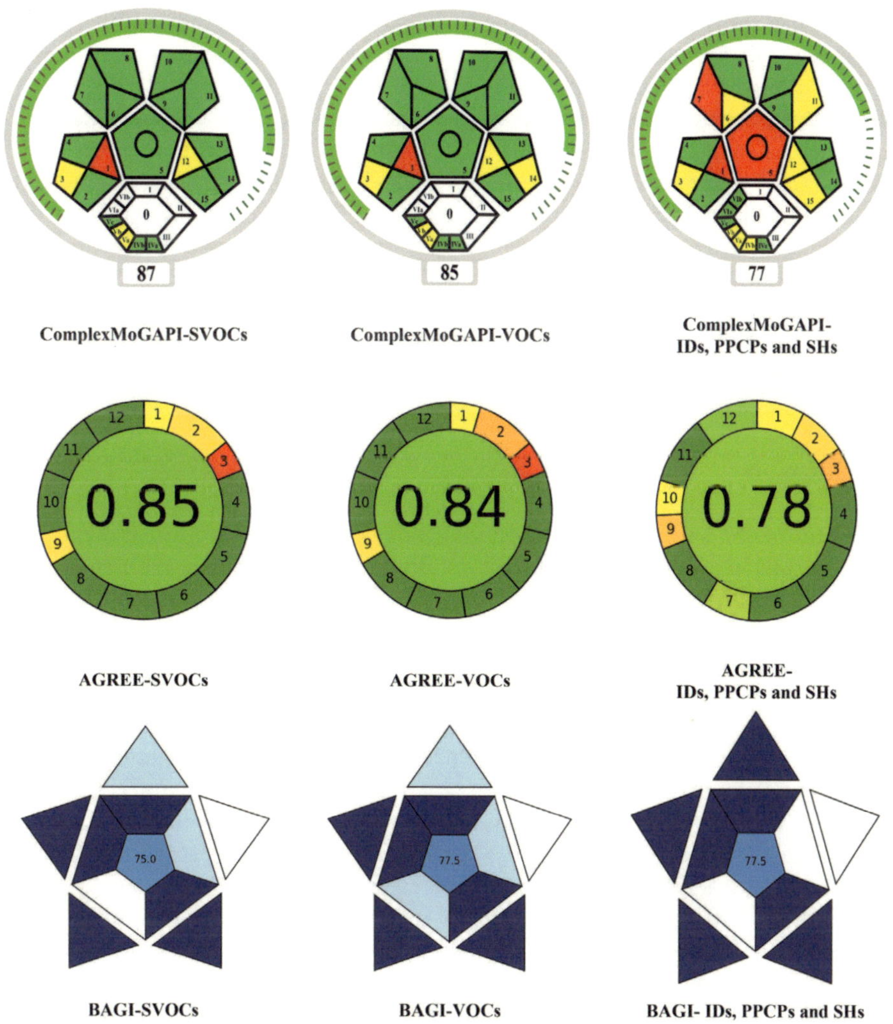


**Figure 4.**
*Evaluation of method greenness and applicability by MoGAPI, AGREE, and BAGI automated methods for analysis of SVOCs, VOCs, IDs, PPCPs, and SHs.*

if less than 50. Combining pentagram and hexagram assessments, ComplexMoGAPI confirmed that the proposed methods for SVOCs, VOCs, and the combined score of IDs, PPCPs, and SHs align with sustainability criteria, achieving an overall excellent eco-friendly score of 87, 85, and 77, respectively (**Figure 4**, https://mostwiedzy.pl/en/justyna-plotka-wasylka,647762-1/complexgapi).

In AGREE, the graph is based on 12 sections representing Green Analytical Chemistry (GAC) principles, each scored from one to zero. A score of 1 indicates a fully green (environmentally friendly) practice, while a score of 0 represents a non-green practice, indicated by red. The overall score is shown in the middle of the pictogram with values close to 1 and a dark green color indicating that the assessed procedure is greener and eco-friendly. As shown in **Figure 4**, for the reported method, eight principles met the method greenness requirement; however, due to offline sample preparation third principle does not meet the requirement of green analysis. However, the proposed method aligns with sustainability criteria and greenness score from AGREE, achieving an overall eco-friendly score of 0.85, 0.84, and 0.78 for the automated method of SVOCs, VOCs, and the combined score of IDs, PPCPs, and SHs, respectively (**Figure 4**, https://mostwiedzy.pl/en/wojciech-wojnowski,174235-1/AGREE).

BAGI evaluates 10 core attributes, such as the type of analysis, the number of analytes assessed in a single method, the simultaneous analysis of multiple analytes, the analytical technique employed, sample analysis rate per hour, level of automation, reagents and materials used, instrument type, sample preparation process, and processing volume. These inputs are used to create an asteroid-shaped pictogram for the proposed method with scoring from 25 to 100, where high score of 75, 77.5, and 77.5 for automated method of SVOCs, VOCs and combine score of IDs, PPCPs and SHs respectively indicates greater practicality of reported method as shown in **Figure 4**, https://mostwiedzy.pl/en/justyna-plotka-wasylka,647762-1/BAGI.

## 4. Conclusion

The automated sample preparation methods of RTC-PAL-SPME Arrow for SVOCs, Headspace-Trap for VOCs, and the Biomek i7 workstation with 96-well SPE (MCX, MAX, and HLB) plates for IDs, PPCPs, and SHs were proven to be fit for the purpose. The integration of high-throughput sample preparation and detection employing the latest models of APGC-MS/MS, GC-MS, and LC-MS/MS has demonstrated the presence of quasi-molecular ions in the MS spectra in most cases. This integration ensures consistency in retention time (RT) and ion ratio, as well as linearity, mean percentage recovery, and precision at the limits of quantification for the most challenging and representative targeted analytes from various chemical classes. The method's suitability for these challenging and representative compounds demonstrates its capability to analyze a wide range of other analytes from different classes for effective WBE. All these methods have the potential to reach further lower levels of LOQs in parts per trillion by increasing the volume of injection and sample. With the switching of the polarity, both negatively and positively ionized molecules were analyzed using the latest models of LC-MS/MS in single injections with satisfactory results for method performance. These automated methods have also achieved high scores in greenness evaluations (GAPI, AGREE) and have received an impressive Blue Applicability Grade Index (BAGI) score for their practicality and environmental sustainability. These eco-friendly, high-throughput methods can be used in labs for various chemical compounds after in-house verification and validation.

## Acknowledgements

Authors duly acknowledge the management of the RASID Laboratory of Abu Dhabi Quality and Conformity Council (ADQCC) and M42 Environmental Sciences, UAE, for providing all resources and facilities to carry out research work.

## Conflict of interest

The authors declare no conflict of interest.

## Author details

Dnyaneshwar Shinde[1*], Bhaskar Karubothula[1], Parth Gupta[1], Raghu Tadala[1], Alanood Almansoori[1], Maryam AlQaydi[1], Samara Bin Salem[2], Wael Elamin[1] and Grzegorz Brudecki[1]

1 RASID Laboratory of Abu Dhabi Quality and Conformity Council (ADQCC) and M42 Environmental Sciences, UAE

2 Abu Dhabi Quality and Conformity Council (ADQCC), UAE

*Address all correspondence to: dshinde@m42.ae

## References

[1] Daughton CG. Pharmaceuticals and personal care products in the environment: Overarching issues and overview. In: ACS Symposium Series. U.S. Government Work. American Chemical Society; 2001. pp. 2-38. Available from: https://pubs.acs.org/doi/10.1021/bk-2001-0791.ch001

[2] Lucci P, Pacetti D, Nunez O, Frega NG. Current trends in sample treatment techniques for environmental and food analysis. In: Chromatography—The Most Versatile Method of Chemical Analysis. ResearchGate; 2012. Available from: https://www.researchgate.net/publication/267508770

[3] Shinde D, Murugan V, Dhagat U, Gupta P, Tadala R, Karubothula B, et al. Introduction of a small volume ethyl acetate based liquid-liquid extraction procedure for analysis of polycyclic aromatic hydrocarbons in wastewater by atmospheric pressure gas chromatography-mass spectrometry and evaluation of method greenness. Journal of Chromatography A. 2025;**1740**:465563. Available from: https://pubmed.ncbi.nlm.nih.gov/39642662/

[4] Method 1694: Pharmaceuticals and Personal Care Products in Water, Soil, Sediment, and Biosolids by HPLC/MS/MS. 2007. Available from: https://www.epa.gov/sites/default/files/201510/documents/method_1694_2007.pdf

[5] Method 1698: Steroids and Hormones in Water, Soil, Sediment, and Biosolids by HRGC/HRMS. 2007. Available from: https://www.epa.gov/sites/default/files/2015-10/documents/method_1698_2007.pdf

[6] Belén F, Gutiérrez J, Springer V. Recent advances in environmental sample preparation: Nanomaterials, automation, and sustainable strategies. Advances in Sample Preparation. 2025;**14**:100171

[7] da Silva S, Burato J, Vargas Medina DA, Toffoli AL, Vasconcelos Soares Maciel E, Mauro LF. Recent advances and trends in miniaturized sample preparation techniques. Journal of Separation Science. 2019;**43**(1):202-225

[8] U.S. EPA. Administration Finalizes Ban on Most Uses of Methylene Chloride, Protecting Workers and Communities from Fatal Exposure. 2024. Available from: https://www.epa.gov/newsreleases/biden-harris-administration-finalizes-ban-most-uses-methylene-chloride-protecting

[9] U.S. EPA (United States Environmental Protection Agency). Method 625.1: Base/Neutrals and Acids by GC/MS. 2016. Available from: https://www.epa.gov/sites/default/files/2017-08/documents/method_625-1_2016.pdf

[10] U.S. EPA (United States Environmental Protection Agency). Method 3510C Separatory Funnel Liquid-Liquid Extraction. 1996. Available from: https://www.epa.gov/sites/default/files/2015-12/documents/3510c.pdf

[11] U.S. EPA (United States Environmental Protection Agency). Method 8100 Polynuclear Aromatic Hydrocarbons. 1986. Available from: https://www.epa.gov/sites/default/files/2015-12/documents/8100.pdf

[12] U.S. EPA (United States Environmental Protection Agency). Method 8410 Gas Chromatography/Fourier Transform Infrared

Spectrometry for Semi Volatile Organics: Capillary Column. 2014. Available from: https://www.epa.gov/sites/default/files/2015-12/documents/8410.pdf

[13] U.S. EPA (United States Environmental Protection Agency). Method 8270E Semi Volatile Organic Compounds BY Gas Chromatography/Mass Spectrometry. 2018. Available from: https://www.epa.gov/sites/default/files/2020-10/documents/method_8270e_update_vi_06-2018_0.pdf

[14] U.S. EPA (United States Environmental Protection Agency). Method 610: Polynuclear Aromatic Hydrocarbons, Appendix A To Part 136 Methods for Organic Chemical Analysis of Municipal and Industrial Wastewater. 1984. Available from: https://www.epa.gov/sites/default/files/2015-10/documents/method_610_1984.pdf

[15] U.S. EPA (United States Environmental Protection Agency). Method 8310 Polynuclear Aromatic Hydrocarbons. 1886. Available from: https://www.epa.gov/sites/default/files/2015-12/documents/8310.pdf

[16] U.S. EPA (United States Environmental Protection Agency). Method 525.3 Determination of Semi Volatile Organic Chemicals in Drinking Water by Solid Phase Extraction and Capillary Column and GC/MS. 2012. Available from: https://cfpub.epa.gov/si/si_public_record_report.cfm?Lab=NERL&dirEntryId=241188

[17] U.S. EPA (United States Environmental Protection Agency). Method 1625C Semi Volatile Organic Compounds by Isotope Dilution GCMS. 1989. Available from: https://www.epa.gov/sites/default/files/2015-09/documents/method_1625c_1989.pdf

[18] APHA (American Public Health Association). Method 6410B, Liquid-Liquid Extraction Gas Chromatographic/Mass Spectrometric Method. 2000. DOI: 10.2105/SMWW.2882.124

[19] International Standard. ISO 7981-2:2005(E) Water Quality—Determination of Polycyclic Aromatic Hydrocarbons (PAH). 2005. Available from: https://www.iso.org/standard/33884.html

[20] International Standard. ISO 17993 Water Quality—Determination of 15 Polycyclic Aromatic Hydrocarbons (PAH) in Water by HPLC with Fluorescence Detection After Liquid-Liquid Extraction. 2002. Available from: https://www.iso.org/standard/31666.html

[21] American Society for Testing and Materials. Standard Test Method for Quantification of Complex Polycyclic Aromatic Hydrocarbon Mixtures or Petroleum Oils in Water. 2024. Available from: https://store.astm.org/d5412-93r24.html

[22] Cherta L, Portolés T, Beltran J, Pitarch E, Mol JGJ, Hernández F. Application of gas chromatography–(triple quadrupole) mass spectrometry with atmospheric pressure chemical ionization for the determination of multiclass pesticides in fruits and vegetables. Journal of Chromatography A. 2013;**1314**:224-240

[23] Pawliszyn J. Solid-phase microextraction in perspective. In: Handbook of Solid Phase Microextraction. Oxford: Elsevier; 2011. pp. 1-12. DOI: 10.1016/B978-0-12-416017-0.00001-2

[24] SANTE 11312/2021 v2 2024. Analytical Quality Control and Method Validation Procedures for Pesticide

Residues Analysis in Food and Feed, EURL for Residues of Pesticides. 2024. Available from: https://food.ec.europa.eu/system/files/2023-11/pesticides_mrl_guidelines_wrkdoc_2021-11312.pdf

[25] U.S. EPA (United States Environmental Protection Agency). Method 624.1: Purgeables by GC/MS. 2016. Available from: https://www.epa.gov/sites/default/files/2017-08/documents/method_624-1_2016.pdf

[26] U.S. EPA (United States Environmental Protection Agency). Method 601: Purgeable Halocarbons. 1994. Available from: https://www.epa.gov/sites/default/files/2015-08/documents/method_601_1994.pdf

[27] U.S. EPA (United States Environmental Protection Agency). Method 8260b Volatile Organic Compounds by Gas Chromatography/Mass Spectrometry (GC/MS). 1996. Available from: https://19january2017snapshot.epa.gov/sites/production/files/2015-12/documents/8260b.pdf

[28] U.S. EPA (United States Environmental Protection Agency). Method 8260c Volatile Organic Compounds by Gas Chromatography/Mass Spectrometry (GC/MS). 2006. Available from: https://19january2017snapshot.epa.gov/sites/production/files/2015-12/documents/8260c.pdf

[29] U.S. EPA (United States Environmental Protection Agency). Method 1624, Revision B: Volatile Organic Compounds by Isotope Dilution GC/MS. 1984. Available from: https://www.epa.gov/sites/default/files/2015-10/documents/method_1624b_1984.pdf

[30] U.S. EPA (United States Environmental Protection Agency). Method 5030c Purge-and-Trap for Aqueous Samples. 2003. Available from: https://www.epa.gov/sites/default/files/2015-07/documents/epa-5030c.pdf

[31] U.S. EPA (United States Environmental Protection Agency). METHOD 524.2. Measurement of Purgeable Organic Compounds in Water by Capillary Column Gas Chromatography/Mass Spectrometry Revision 4.1. 1995. Available from: https://www.epa.gov/sites/default/files/2015-06/documents/epa-524.2.pdf

[32] International Standard. ISO 10301:1997 Water Quality-Determination of Highly Volatile Halogenated Hydrocarbons-Gas-Chromatographic Methods. 1997. Available from: https://www.iso.org/standard/18345.html

[33] Official Journal of the European Union. Directive (EU) 2020/2184 of the European Parliament and of the Council of 16 December 2020 on the Quality of Water Intended for Human Consumption. Available from: https://eur-lex.europa.eu/legal-content/EN/TXT/PDF/?uri=CELEX:32020L2184

[34] Asimakopoulos AG, Kannan P, Higgins S, Kannan K. Determination of 89 drugs and other micropollutants in unfiltered wastewater and freshwater by LC-MS/MS: An alternative sample preparation approach. Analytical and Bioanalytical Chemistry. 2017;**409**(26):6205-6225

[35] Centazzo N, Frederick BM, Jacox A, Cheng SY, Concheiro-Guisan M. Wastewater analysis for nicotine, cocaine, amphetamines, opioids and cannabis in New York City. Forensic Sciences Research. 2019;**4**(2):152-167

[36] Sulej-Suchomska AM, Klupczynska A, Dereziński P, et al. Urban wastewater analysis as an effective tool for monitoring illegal drugs,

including new psychoactive substances, in the Eastern European region. Scientific Reports. 2020;**10**:4885. DOI: 10.1038/s41598-020-61628-5

[37] Senta I, Krizman I, Ahel M, Terzić S. Multiresidual analysis of emerging amphetamine-like psychoactive substances in wastewater and river water. Journal of Chromatography A. 2015;**1425**:204-212

[38] Gul W, Stamper B, Godfrey M, Gul SW, ElSohly MA. LC–MS-MS method for analysis of opiates in wastewater during football games II. Journal of Analytical Toxicology. 2016;**40**(5):330-337

[39] Kim U, Sim W, Cha H, Oh J-E. Development of a simultaneous multi–component analytical method for 37 new psychoactive substances in wastewater using LC–MS/MS. Environmental Engineering Research. 2023;**29**. DOI: 10.4491/eer.2023.491

[40] Gracia-Marín E, Hernández F, Ibáñez M, Bijlsma L. Dilute-and-shoot approach for the high-throughput LC-MS/MS determination of illicit drugs in the field of wastewater-based epidemiology. Water Research. 2024;**259**:121864

[41] Yuan S, Xiang Y, Chen L, Xiang P, Li Y. Magnetic solid-phase extraction based on polydopamine-coated magnetic nanoparticles for rapid and sensitive analysis of eleven illicit drugs and metabolites in wastewater with the aid of UHPLC-MS/MS. Journal of Chromatography A. 2024;**1718**:464703

[42] Liu Y, Cao S, Tian J, You J, Liu Z, Chen Z. A magnetic solid phase extraction based on DES/ZIF-MGO coupled with UPLC-MS/MS for the simultaneous detection and consumption evaluation of four illicit drugs. Microchemical Journal. 2024;**200**:110448

[43] Li K, Hu Y, Jiang Y, Han X, Liu X, Du M. Direct-injection UHPLC-MS/MS method for simultaneous determination of 78 illegal drugs and psychoactive substances in domestic wastewater. Water. 2024;**16**(9):1315-1315

[44] Nadarajan D, O'Brien J, Cresswell S, Kele B, Mueller J, Bade R. Application of design of experiment for quantification of 71 new psychoactive substances in influent wastewater. Analytica Chimica Acta. 2024;**1321**:343036

[45] Ortúzar M, Esterhuizen M, Olicón-Hernández DR, González-López J, Aranda E. Pharmaceutical pollution in aquatic environments: A concise review of environmental impacts and bioremediation systems. Frontiers in Microbiology. 2022;**13**:869332. DOI: 10.3389/fmicb.2022.869332

[46] Omuferen LO, Maseko B, Olowoyo JO. Occurrence of antibiotics in wastewater from hospital and convectional wastewater treatment plants and their impact on the effluent receiving rivers: Current knowledge between 2010 and 2019. Environmental Monitoring and Assessment. Mar 2022. DOI: 10.1007/s10661-022-09846-4

[47] Murray CJL. Global Burden of Bacterial Antimicrobial Resistance 1990-2021: A Systematic Analysis with Forecasts to 2050. Elsevier; 2024. DOI: 10.1016/S0140-6736(24)01867-1

[48] US EPA SDWA. Contaminant Candidate List 4-CCL 4. 2021. Available from: https://www.epa.gov/ccl/contaminant-candidate-list-4-ccl-4-0

[49] U.S. EPA SDWA. Contaminant Candidate List 3—CCL 3. 2014. Available from: https://www.epa.gov/

ccl/contaminant-candidate-list-3-ccl-3#ccl3-list

[50] Benedé JL, Cagliero C, Nemutlu E, Pena-Pereira F, Bicchi C, Carrasco-Correa EJ, et al. Greenness assessment of 174 CEN, ISO, and pharmacopoeia standard methods and their sub-methods used for environmental, food, trace element and pharmaceutical analyses. Advances in Sample Preparation. 2025;**14**:100180. DOI: 10.1016/j.sampre.2025.100180. ISSN 2772-5820

[51] López-Lorente ÁI, Pena-Pereira F, Pedersen-Bjergaard S, Zuin VG, Ozkan SA, Psillakis E. The ten principles of green sample preparation. TrAC Trends in Analytical Chemistry. 2022;**148**:116530

[52] Wojnowski W, Tobiszewski M, Pena-Pereira F, Psillakis E. AGREEprep—Analytical greenness metric for sample preparation. TrAC Trends in Analytical Chemistry. 2022;**149**:116553

[53] Płotka-Wasylka J, Wojnowski W. Complementary green analytical procedure index (ComplexGAPI) and software. Green Chemistry. 2021;**23**(21):8657-8665

[54] Manousi N, Wojnowski W, Płotka-Wasylka J, Samanidou V. Blue applicability grade index (BAGI) and software: A new tool for the evaluation of method practicality. Green Chemistry. 2023;**25**(19):7598-7604

Chapter 2

# High-Throughput Glycomics Analysis Enabled by Multiplex Chemical Derivatization and Mass Spectrometry

*Miyang Li*


## Abstract

Glycomics is dedicated to elucidating the structure and function of glycome, which constitutes the full suite of glycans produced by a cell or organism, as well as the regulatory principles behind it. This formidable challenge necessitates the development of robust, sensitive, and scalable technologies for the rapid analysis of numerous glycans, driving the field of high-throughput glycomics. While current methodologies, such as U(H)PLC-FLD, CGE-LIF, and MALDI/LC-MS, have greatly enabled glycomics study, they are often limited by serial sample preparation and sequential analysis, which constrains throughput and introduces inter-batch variability and quantitative inaccuracy. This chapter highlights a critical technological advancement that utilizes the multiplexed isobaric labeling strategy (e.g., aminoxyTMT, SUGAR tags) for analyzing multiple samples by LC-MS/MS at one time. We discuss how this approach dramatically enhances throughput, reduces analytical time and cost, improves data quality by minimizing batch effects, and enables precise relative quantification across sample sets. By focusing on the principles, applications, and future potential of multiplexed glycomics, this chapter underscores its transformative role in advancing glycan profiling for extensive biopharmaceutical studies.


**Keywords:** glycomics, chemical derivatization, isobaric tags, relative quantification, mass spectrometry

## 1. Introduction

Glycosylation, the enzymatic attachment of carbohydrates to proteins and lipids, is a ubiquitous and critically important post-translational modification (PTM). It generates a remarkable diversity of glycoconjugates (**Figure 1**), including glycoproteins, proteoglycans, and glycosphingolipids (GSLs), alongside other glycan classes such as free oligosaccharides, glycosaminoglycans (GAGs), and the recently discovered glycosylated RNAs [1]. This structural diversity underpins a vast array of biological functions, influencing protein folding, stability, solubility, immunogenicity, and cellular communication [2]. Given its central role, aberrant glycosylation is

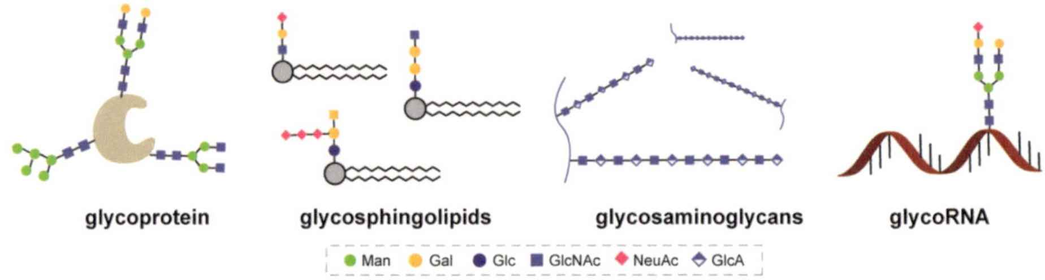


**Figure 1.**
*Schematic representation of major glycan classes, including glycoproteins, glycosphingolipids, proteoglycans/glycosaminoglycans (GAGs), and the recently identified glycoRNAs.*

a hallmark of numerous diseases, including cancer, congenital disorders of glycosylation, and autoimmune diseases [3]. Consequently, the comprehensive analysis of the glycome—the complete set of glycans in a biological system—has become a major objective in biomedical research, driving the need for high-throughput analytical methods to decipher glycan functions and discover glycan-based biomarkers in large cohort studies.

Significant progress has been made in high-throughput glycomics, as comprehensively reviewed elsewhere [4]. Core methodologies typically involve the release of glycans (e.g., *via* PNGase F digestion for N-glycans), followed by separation-based techniques like ultra-high-performance liquid chromatography with fluorescence detection (UHPLC-FLD) or capillary gel electrophoresis with laser-induced fluorescence (CGE-LIF), and liquid chromatography-mass spectrometric (LC-MS) analysis. While these platforms have been successfully applied to profile research and clinical samples, they operate predominantly in a serial analysis mode. This presents inherent limitations: analysis time scales linearly with the number of samples, instrument duty cycle is underutilized, and technical variability between processing batches can introduce significant noise, complicating data integration and interpretation in large-scale studies.

A paradigm-shifting solution to these bottlenecks is the adoption of multiplexed isobaric labeling. This approach enables researchers to covalently tag the reducing terminus of glycans from different samples with unique isobaric tags, such as aminoxy tandem mass tags (aminoxyTMT) or isobaric tags for relative and absolute quantitation of glycans (SUGAR). These tags confer identical precursor masses to labeled glycans but yield distinct, quantifiable reporter ions upon collisional dissociation in MS/MS. Consequently, glycans from multiple different samples (currently up to 12-plex) can be pooled and analyzed in a single LC-MS/MS injection. The relative abundance of each glycan across the original samples is then determined by comparing the intensities of the respective reporter ions, dramatically increasing throughput while reducing technical variance and instrument time per sample.

This chapter addresses a critical gap in the literature by focusing on this powerful yet underappreciated strategy. We will elucidate the chemistry behind leading multiplex tags, detail their integration into analysis workflows, and discuss their profound impact on enhancing throughput, quantitative precision, and data quality in high-throughput glycomics. By highlighting the advantages and applications of multiplexed glycan analysis *via* isobaric labeling, we position it as a foundational technique for the next generation of large-scale glycomics studies, poised to accelerate discovery in personalized medicine and biotherapeutic development.

## 2. Historical overview and methodological evolution in glycomics

Among the diverse array of glycoconjugates, glycoproteins represent a class of exceptional complexity and functional significance. They are primarily classified by their glycosidic linkage: N-linked glycans (attached to the asparagine residue of the canonical Asn-X-Ser/Thr sequon, where X is any amino acid except proline) and O-linked glycans (attached to serine or threonine without a strict consensus sequence). This structural complexity is further defined by two levels of heterogeneity: macroheterogeneity, referring to the occupancy and selection of glycosylation sites on a polypeptide, and microheterogeneity, describing the diverse repertoire of glycan structures that may occupy a single site. Critically, the specific glycoform expressed is not directly genetically encoded but is dynamically regulated by cellular status, including protein conformation, the availability of glycosyltransferases and nucleotide sugar donors, and metabolic flux [5].

Deciphering this heterogeneity is essential for understanding biological function and is particularly crucial in the biopharmaceutical development, where glycosylation directly impacts drug efficacy, stability, and immunogenicity. The analytical strategy employed is dictated by the research question, leading to a hierarchy of methodologies. Intact mass analysis provides the most comprehensive view by characterizing proteoforms with their full complement of PTMs. However, it requires highly purified samples, advanced high-resolution mass spectrometry, and complex data deconvolution, rendering it impractical for high-throughput screens. Middle-down approaches, often used for antibodies (e.g., enzymatic cleavage into Fc/Fab domains), simplify intact analysis to better resolve macroheterogeneity but still struggle with the analytical challenges posed by extensive microheterogeneity [6, 7].

In contrast, bottom-up glycoproteomics involves proteolytic digestion followed by LC-MS/MS to analyze glycopeptides. This powerful approach provides site-specific information on both occupancy (macroheterogeneity) and the attached glycan structures (microheterogeneity). Nevertheless, the sheer structural diversity at each site can create significant analytical complexity, often acting as a bottleneck for throughput. This is frequently mitigated by targeted affinity enrichment strategies for specific glycoproteins or glycopeptides [8].

For large-scale epidemiological and clinical biomarker discovery studies, the analysis of released glycans has emerged as the foremost high-throughput strategy (**Figure 2**). This approach involves the enzymatic or chemical liberation of glycans from their protein scaffolds, enabling the profiling of the global glycome from complex biological matrices without prior protein purification. The release of N-glycans is efficiently achieved with the universal enzyme peptide-N-glycosidase F (PNGase F). The analysis of O-glycans remains more challenging, often relying on chemical methods like reductive β-elimination due to the lack of a broad-specificity enzyme. Since native glycans lack efficient chromophores or fluorophores, they typically require chemical derivatization [9]. Early foundational—though now largely obsolete—methods involved hazardous techniques like radioactive labeling coupled with hydrazinolysis for release, a process hampered by the use of explosive anhydrous hydrazine [10].

A pivotal advancement was the widespread adoption of fluorescent tagging *via* reductive amination. First described by Hase et al. in 1979 for free oligosaccharides [11], this method was coupled with HPLC-fluorescence detection (HPLC-FLD) to achieve vastly improved sensitivity. The reaction involves conjugating the reducing end of the glycan to an amine-containing fluorophore—such as 2-aminobenzamide (2-AB),

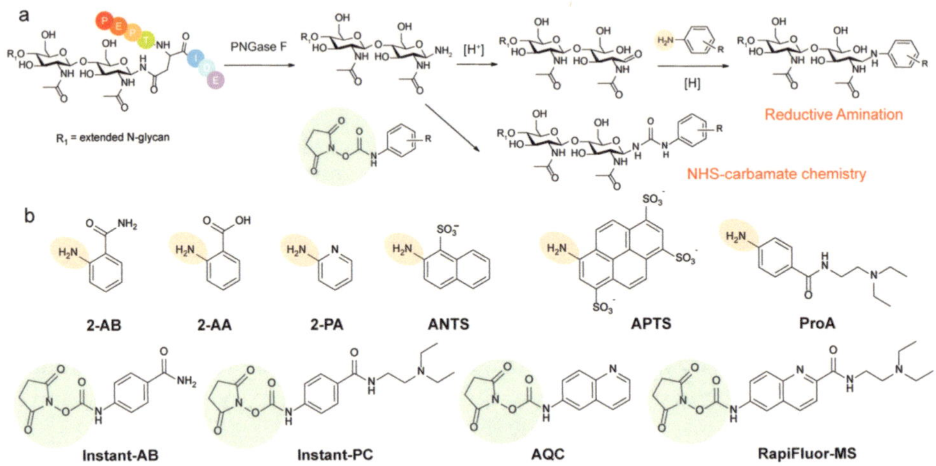


**Figure 2.**
*Evolution of released glycan analysis. (a) Standard workflow for N-glycan release* via *PNGase F and subsequent derivatization at the reducing end. (b) Common fluorophores used in glycan reducing-end labeling.*

2-aminopyridine (2-AP), or procainamide (ProA) [12]—using a reducing agent like sodium cyanoborohydride ($NaBH_3CN$) to create a stable derivative. This paradigm enabled sensitive detection from the picomole amounts of material [13] and established the core workflow for high-throughput glycomics.

A more recent innovation aims to bypass the slower reductive amination step by targeting the reactive glycosylamine group immediately after enzymatic release with PNGase F. This approach facilitates rapid labeling with tags such as RapiFluor-MS [14], InstantPC [15], and DiPyrO [16], significantly reducing sample handling time and improving efficiency.

Thus, the historical trajectory of glycomics has been defined by the paradigm of enzymatic release (greatly enabled by PNGase F), chemical derivatization (evolving from hazardous procedures to efficient fluorescent tagging), and chromatographic or electrophoretic separation. This established the robust foundation upon which modern high-throughput glycomics is built, setting the stage for the next revolutionary advance: multiplexed isobaric labeling.

## 3. Isobaric labeling techniques for high-throughput glycomics

The strategic power of isobaric labeling was first realized in proteomics, where tags like Tandem Mass Tags (TMT) [17] and Isobaric tags for relative and absolute quantitation (iTRAQ) [18] revolutionized quantitative analysis by enabling the simultaneous comparison of multiple samples in a single LC-MS/MS run. This approach minimizes missing values and enhances statistical confidence by reducing run-to-run variability. The fundamental design of these tags involves three key components: [1] a reactive group that targets specific functional groups on analytes (e.g., amines on peptides); [2] a reporter moiety that, upon MS/MS fragmentation, yields a unique low-mass ion for quantification; and [3] a mass balancer that ensures the total mass of the tag is identical (isobaric) across all versions, allowing co-elution and combined MS1 detection [19].

The success in proteomics provided a clear blueprint for adapting this strategy to glycomics. By designing tags that target the reducing end of glycans, researchers have integrated multiplexing into the standard released glycan workflow, creating a powerful paradigm for high-throughput analysis (**Figure 3**). This section details the key isobaric and isotopic tags developed for this purpose (**Figure 4**).

### 3.1 INLIGHT™

Although not a true isobaric multiplexing technique, INLIGHT (Individuality Normalization when Labeling with Isotopic Glycan Hydrazide Tags) was a foundational advancement in MS-compatible, isotope-coded glycan labeling [20]. This strategy employs a pair of stable isotope-labeled hydrazide tags, enabling duplex quantification based on the area under the curve (AUC) of precursor ions in full MS scans, analogous to the SILAC method in proteomics.

The INLIGHT tags were rationally designed to address the poor ionization efficiency of native glycans. The optimal tag features a hydrophobic biphenyl structure that mitigates ionization bias and enhances ESI-MS sensitivity [21, 22]. A key innovation of the INLIGHT strategy is its sophisticated data processing, which includes

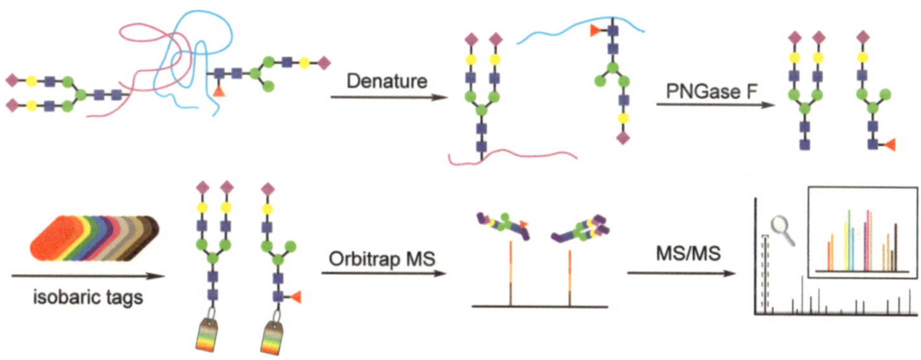


**Figure 3.**
*Workflow for isobaric labeling-enabled high-throughput glycomics. The process involves releasing glycans from individual samples, labeling each pool with a unique isobaric tag, combining the samples, and analyzing them in a single LC-MS/MS run. Quantification is achieved by measuring the intensity of reporter ions released during the MS/MS fragmentation.*

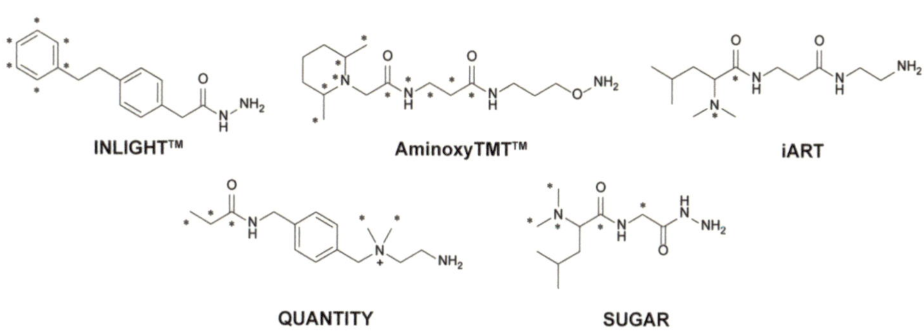


**Figure 4.**
*Chemical structures of isobaric/isotopic tags developed for glycomics analysis. (A star denotes the position of isotopes used to create isobaric sets or light/heavy pairs.)*

correction for isotopic overlap in high-mass glycans and a post-acquisition normalization method (Total Glycan Normalization Factor, TGNF) to correct for systematic variability more effectively than a single internal standard.

Commercialized by Cambridge Isotope Laboratories, Inc. and further developed, INLIGHT has been adapted for O-glycan and heparin oligomer analysis [23, 24], with evolving protocols and dedicated software [25–28]. Its applications have provided insights into areas such as xylem differentiation and ovarian cancer biomarker discovery [26, 29]. More recently, coupling INLIGHT with ion mobility spectrometry (IMS) has facilitated rapid glycan subclass identification based on molecular trend lines defined by m/z and collision cross-section (CCS) values [30].

### 3.2 AminoxyTMT™

A significant milestone was reached with the development of carbonyl-reactive tandem mass tags (glyco-TMTs) by Kuhn et al. in 2012, marking the first true isobaric reagents for glycans [31]. These tags adapted the proven TMT scaffold, testing both hydrazide and aminooxy functionalities as the reactive group. The aminooxy group was found to be superior, facilitating facile labeling of the glycan reducing end and demonstrating exceptional performance: labeling efficiencies greater than 95%, a coefficient of variation below 10%, and a dynamic range extending up to a 1:40 ratio. Such capabilities squarely address the critical need for robust quantification in glycomics research.

It is noteworthy that the initial study created not only duplex isobaric channels but also a light/heavy isotopic version, similar in principle to INLIGHT. This design allowed for a direct comparison of quantification methods within the same platform. The authors evaluated reporter ion quantification (the core isobaric tagging principle) against full MS quantification based on the area under the curve of the precursor ion. They concluded that under their experimental conditions, reporter ion quantification was less superior, primarily due to the inherent limitations of the tag chemistry discussed below.

Commercialized by Thermo Fisher Scientific as 6-plex AminoxyTMT, it remains the only commercially available isobaric tag specifically designed for glycan analysis. Its early application to N-glycan profiles from isogenic colorectal cancer cell lines SW480 and SW620 revealed significant differences between primary and metastatic tumors, underscoring its biomedical potential. Subsequent studies have leveraged it to identify differentiated glycan profiles in blood serum from patients with esophageal diseases [32], and dedicated software like SimGlycan has been developed to streamline data processing [33].

However, AminoxyTMT faces notable limitations that explain the performance gap observed in the original study. A primary challenge is the inefficiency of reporter ion generation. The TMT tag was originally optimized for peptide fragmentation, where the energy required to release the reporter ion matches that needed to break amide bonds. This energy is significantly higher than that required to fragment glycosidic bonds. Consequently, during MS/MS of glycans, energy is diverted into extensive carbohydrate fragmentation, leading to reduced recovery of the TMT reporter ions. Furthermore, the oxime bond formed by the aminooxy group is more labile than the stable secondary amine formed in reductive amination, increasing the probability of entire tag loss, which further suppresses reporter ion signal. The resulting low-reporter ion intensity and poor signal-to-noise ratio inevitably compromise quantification accuracy and dynamic range. Another practical issue is the interference of the 126 Da reporter ion with a common HexNAc fragment (Δm = 0.07 Da), which cannot be resolved on medium-resolution instruments, effectively reducing the practical multiplexing capacity [31].

### 3.3 iART

Developed by Zhou et al. in 2013, isobaric aldehyde reactive tags (iARTs) represent an alternative strategy that leverages the well-established chemistry of reductive amination for glycan derivatization [34]. The hydrophobic nature of the iART structure enhances ionization efficiency in ESI-MS, yielding stronger signals even at lower glycan concentrations. The authors rigorously validated the quantitative accuracy of iARTs by demonstrating a strong consistency between the observed and expected ratios in labeled oligosaccharide standards, underscoring the reliability of the method for glycomic studies.

The practical utility of iARTs was demonstrated in disease research through the analysis of glycans from HIV gp120 glycoproteins, providing valuable insights into glycosylation patterns critical for understanding immune responses. Furthermore, the study highlighted the potential for expanding the iART technology to a six-plex set, which would enable simultaneous analysis of multiple samples and significantly improve throughput.

However, the study also acknowledged several limitations. A notable challenge is the requirement for higher collision energy to achieve complete fragmentation of larger glycans, which can compromise quantification accuracy. This is evidenced by higher coefficients of variation (CV) reported for glycans exceeding 2000 Da, indicating the reproducibility challenges for complex glycan structures. Additionally, as with any method relying on standard reductive amination, the labeling process under acidic conditions may lead to the loss of acid-labile sialic acids, potentially limiting the method's applicability for profiling sialylated glycans. Another limitation is the actual demonstrated throughput; while a six-plex expansion was proposed, the study only synthesized and practically validated a duplex set (channels 114 and 115). The variability in optimal fragmentation energy across different glycan sizes further complicates the method, necessitating detailed optimization for complex samples.

### 3.4 QUANTITY

As a successor to iART from the same research group, the QUANTITY method introduces a significant chemical innovation by incorporating a quaternary ammonium group into the isobaric tag [35]. This permanent positive charge addresses key challenges faced by its predecessor. It ensures enhanced ionization efficiency and detection sensitivity during MS analysis, making it particularly suitable for analyzing low-abundance glycans or limited sample quantities. Unlike methods that rely on metal ion adducts, QUANTITY avoids associated ion suppression issues. Most importantly, the permanent charge promotes efficient fragmentation during MS/MS, generating strong reporter ions for accurate quantification.

The study demonstrated the versatility of QUANTITY by achieving high labeling efficiency and applying it to the glycomic profiling of complex biological samples, including human serum and genetically engineered Chinese Hamster Ovary (CHO) cells. The method showed the excellent reproducibility and quantitative accuracy, with a strong linear correlation between theoretical and experimental ratios, underscoring its robustness.

One of the major strengths of QUANTITY is its seamless integration into existing workflows; it is compatible with the established protocols for 2-AA/2-AB labeling, allowing for easier adoption. However, the method has some limitations. Its multiplexing capacity is currently limited to four samples, which is lower than the commercially available 6-plex AminoxyTMT and may restrict its utility in very large-scale studies. Additionally, the reported protocol involves specialized steps, such as solid-phase glycoprotein immobilization and the use of specific reagents like *p*-toluidine for stabilizing sialic acids. These requirements could pose implementation challenges for laboratories lacking the necessary materials or technical expertise.

### 3.5 SUGAR

The SUGAR (Isobaric Multiplex Labeling Reagents for Carbonyl-Containing Compound) tags were developed as an economic and synthetically accessible alternative to TMT, building upon the compact dimethylated leucine (DiLeu) scaffold originally created for quantitative proteomics [36]. The structural simplicity of the DiLeu tag allows for straightforward synthetic extension, leading to a versatile family of reagents such as DiLeuPMP for O-glycan analysis [37], and DiLeuEN for glycopeptide and lipid analysis [38]. SUGAR tags employ a hydrazide functionality to target the glycan reducing end. Initially developed as a 4-plex set, the platform has been significantly expanded to a 12-plex format, representing the highest multiplexing capacity available for glycomics to date [39, 40].

The application of SUGAR tags has proven valuable in clinical and cell biology research. For instance, their use revealed that most N-glycans were down-regulated in B-cells from pediatric acute lymphoblastic leukemia patients after chemotherapy. A similar trend was observed in endometrial cancer cells treated with a quinone antimicrobial medication, highlighting the sensitivity of glycosylation as a biomarker for drug response. Furthermore, the platform's flexibility is demonstrated by its combination with a NeuCode strategy, which creates mass-defect versions of the tags. This allows for targeted analysis of low-abundance glycans by leveraging ultra-high-resolution mass spectrometers to differentiate mini-dalton mass differences [41, 42].

A critical performance metric for any isobaric tag is the efficiency of reporter ion generation. While the study demonstrated that SUGAR tags generate reporter ions more efficiently than AminoxyTMT, the intensity of these reporter ions was reported to be only around 10–20% of the most abundant oxonium ions (e.g., *m/z* 204 from HexNAc or *m/z* 366 from HexNAc-Man). This relative inefficiency can impact quantification sensitivity. Moreover, the low mass of the SUGAR reporter ions (below *m/z* 120) presents an analytical constraint: It necessitates setting a first mass cutoff during MS/MS method development. This setting prevents the detection of high-mass fragments that are crucial for glycan structural elucidation, potentially complicating data interpretation.

## 4. Limitations of current methods

The integration of isobaric labeling has undeniably accelerated high-throughput glycomics, creating a powerful synergy between novel probe chemistry, efficient derivatization, and advanced LC-MS/MS. As summarized in **Table 1**, the development of various tags has provided researchers with a toolkit for multiplexed analysis. Despite these significant advancements, a critical examination reveals several common limitations that represent both current challenges and opportunities for future innovation.

| Tag name | Reactive chemistry | Multiplexing capacity | Key feature(s) | Primary limitation(s) |
|---|---|---|---|---|
| INLIGHT™ | Hydrazide (reductive amination) | 2-Plex (isotopic) | Pioneering isotopic tag; TGNF normalization; enhanced ionization | Not true isobaric multiplexing; limited to duplex analysis. |
| AminoxyTMT™ | Aminooxy (oxime ligation) | 6-Plex | First commercial isobaric tag; facile labelling and high labelling efficiency | Inefficient reporter ion generation; oxime bond lability; reporter ion interference (126 Da). |
| iART | Primary amine (reductive amination) | 2-Plex (proposed 6-plex) | Hydrophobic tag improves ionization | High CV for large glycans; limited to duplex analysis (no follow-up to improve throughput) |
| QUANTITY | Primary amine (reductive amination) | 4-Plex | Permanent charge enhances sensitivity and fragmentation | Limited throughput; protocol requires specialized steps (e.g., solid-phase immobilization). |
| SUGAR | Hydrazide (reductive amination) | 4-Plex 12-Plex | Highest multiplexing capacity; based on synthetically accessible DiLeu scaffold | Low reporter ion intensity relative to oxonium ions; low first mass cutoff masks key structural ions. |

**Table 1.**
*Key characteristics of current isobaric and isotopic labeling platforms for glycomics, highlighting their respective advantages and limitations.*

A fundamental constraint of current isobaric tags is their universal reliance on derivatizing the reducing end of glycans. Most tags, including iART, QUANTITY, and SUGAR, utilize reductive amination. This reaction requires acidic conditions and elevated temperature that are detrimental to acid-labile residues, notably sialic acids, leading to their loss and inaccurate profiling of sialylation [43]. Furthermore, the reaction produces salt byproducts that necessitate the extensive sample cleanup, complicating automation. While AminoxyTMT avoids reductive amination by forming an oxime bond, it introduces a different challenge. The reaction with the glycosylamine intermediate is fast, but if targeting the open-ring reducing end, it first requires a hydrolysis step to expose the reactive carbonyl group. This additional, often time-varying, hydrolysis can introduce analytical variance and lengthen sample preparation time.

As highlighted in the discussion of individual tags, a pervasive issue across all platforms is the inefficient generation of reporter ions. This stems from a fundamental mismatch in bond dissociation energies. The energy required to cleave the tag and release the reporter ion is typically optimized for peptide amide bonds. However, the glycosidic bonds of the glycan itself fragment at lower energies. Consequently, during MS/MS, collision energy is preferentially dissipated through extensive glycan fragmentation, producing dominant oxonium ions (often with further dehydration). The reporter ions are thus overshadowed, leading to lower signal-to-noise ratios. This directly translates to higher quantitative variance and a narrowed effective dynamic range compared to theoretical expectations.

Another gap could be filled in the further is the lack of dedicated, user-friendly bioinformatics platforms for data processing and interpretation. While the wet-lab workflow achieves high throughput, data analysis frequently becomes a manual, time-consuming bottleneck. Unlike in proteomics, where robust, integrated software suites seamlessly handle TMT or iTRAQ data from raw file to final quantitative ratios, the bioinformatics for isobaric glycan analysis often relies on a fragmented combination of proprietary instrument software, in-house scripts, and manual validation. This is compounded by the unique complexities of glycan data, including isobaric overlaps between different compositions, the presence of multiple charge states and adducts, and—crucially—the need to correctly associate MS2 spectra containing both quantitative reporter ions and structural carbohydrate fragments. The low intensity of reporter ions further exacerbates the challenge, requiring the sophisticated algorithms for signal extraction and noise filtering. However, integrated software solutions, such as PEAKS GlycanFinder, are continually evolving to address this technical gap [44].

## 5. Conclusion and future outlook

A significant opportunity for future development lies in bridging the gap between modern MS-based multiplexing and the established sensitivity of fluorescence detection. Classical high-throughput glycomics still heavily relies on fluorescent tags (e.g., 2-AB) coupled with UHPLC-FLD, which offers exceptional sensitivity and robustness against matrix effects. The current isobaric tags forego this capability.

The next generation of tags could be designed to incorporate a fluorophore alongside the isobaric mass tag. This would create a dual-detection platform where the same pooled sample could be analyzed first by UHPLC-FLD for highly sensitive, matrix-resistant quantification, and subsequently by LC-MS/MS for multiplexed, sample-specific quantification and structural confirmation. This orthogonal approach would leverage the strengths of both techniques, potentially boosting overall sensitivity, reliability, and data richness.

In conclusion, while isobaric labeling has propelled glycomics into a new era of throughput, addressing these limitations—through milder chemistries, optimized fragmentation schemes, and multi-modal tags—will be crucial for unlocking its full potential. The ongoing evolution of these tags will continue to deepen our understanding of the glycome in health and disease.

## Author details

Miyang Li
Regeneron Pharmaceuticals, Inc., Tarrytown, NY, United States

*Address all correspondence to: miyangli88@gmail.com

## References

[1] Flynn RA, Pedram K, Malaker SA, Batista PJ, Smith BAH, Johnson AG, et al. Small RNAs are modified with N-glycans and displayed on the surface of living cells. Cell. 2021;**184**(12):3109-24.e22

[2] He M, Zhou X, Wang X. Glycosylation: Mechanisms, biological functions and clinical implications. Signal Transduction and Targeted Therapy. 2024;**9**(1):194

[3] Seeling M, Brückner C, Nimmerjahn F. Differential antibody glycosylation in autoimmunity: Sweet biomarker or modulator of disease activity? Nature Reviews Rheumatology. 2017;**13**(10):621-630

[4] Trbojević-Akmačić I, Lageveen-Kammeijer GSM, Heijs B, Petrović T, Deriš H, Wuhrer M, et al. High-throughput glycomic methods. Chemical Reviews. 2022;**122**(20):15865-15913

[5] Varki A, Cummings RD, Esko JD, Stanley P, Hart GW, Aebi M, et al. Essentials of Glycobiology. Cold Spring Harbor, NY: Cold Spring Harbor Laboratory Press; 2022. ISBN-13: 978-1-621824-21-3 (hardback); ISBN-13: 978-1-621824-22-0 (epub)

[6] Senini I, Tengattini S, Rinaldi F, Massolini G, Gstöttner C, Reusch D, et al. Direct glycosylation analysis of intact monoclonal antibodies combining ESI MS of glycoforms and MALDI-in source decay MS of glycan fragments. Communications Chemistry. 2024;7(1):203

[7] Yang Y, Liu F, Franc V, Halim LA, Schellekens H, Heck AJ. Hybrid mass spectrometry approaches in glycoprotein analysis and their usage in scoring biosimilarity. Nature Communications. 2016;7:13397

[8] Malaker SA. Glycoproteomics: Charting new territory in mass spectrometry and glycobiology. Journal of Mass Spectrometry. 2024;**59**(6):e5034

[9] Li M, Shen S, Letarte S, Flick T. Discovery and characterization of novel G0 glycan isomers in an afucosylated therapeutic antibody using liquid chromatography-mass spectrometry. Journal of the American Society for Mass Spectrometry. 2025;**36**(5):1034-1040

[10] Takasaki S, Kobata A. Microdetermination of individual neutral and amino sugars and N-acetylneuraminic acid in complex saccharides. Journal of Biochemistry. 1974;**76**(4):783-789

[11] Hase S, Ikenaka T, Matsushima Y. Analyses of oligosaccharides by tagging the reducing end with a fluorescent compound. I. Application to glycoproteins. Journal of Biochemistry. 1979;**85**(4):989-994

[12] Váradi C, Hajdu V, Farkas F, Gilányi I, Oláh C, Viskolcz B. The analysis of human serum N-glycosylation in patients with primary and metastatic brain tumors. Life (Basel). 2021;**11**(1):29. Available from: https://www.mdpi.com/2075-1729/11/1/29

[13] Hase S, Sugimoto T, Takemoto H, Ikenaka T, Schmid K. The structure of sugar chains of Japanese quail ovomucoid. The occurrence of oligosaccharides not expected from the classical biosynthetic pathway for N-glycans; A method for the assessment of the structure of glycans present in picomolar amounts. Journal of Biochemistry. 1986;**99**(6):1725-1733

[14] Lauber MA, Yu YQ, Brousmiche DW, Hua Z, Koza SM, Magnelli P, et al. Rapid

preparation of released N-glycans for HILIC analysis using a labeling reagent that facilitates sensitive fluorescence and ESI-MS detection. Analytical Chemistry. 2015;**87**(10):5401-5409

[15] Segu Z, Stone T, Berdugo C, Roberts A, Doud E, Li Y. A rapid method for relative quantification of. MAbs. 2020;**12**(1):1750794

[16] Frost DC, Buchberger AR, Li L. Mass defect-based dimethyl pyrimidinyl ornithine (DiPyrO) tags for multiplex quantitative proteomics. Analytical Chemistry. 2017;**89**(20):10798-10805

[17] Thompson A, Schäfer J, Kuhn K, Kienle S, Schwarz J, Schmidt G, et al. Tandem mass tags: A novel quantification strategy for comparative analysis of complex protein mixtures by MS/MS. Analytical Chemistry. 2003;**75**(8):1895-1904

[18] Ross PL, Huang YN, Marchese JN, Williamson B, Parker K, Hattan S, et al. Multiplexed protein quantitation in Saccharomyces cerevisiae using amine-reactive isobaric tagging reagents. Molecular & Cellular Proteomics. 2004;**3**(12):1154-1169

[19] Li J, Cai Z, Bomgarden RD, Pike I, Kuhn K, Rogers JC, et al. TMTpro-18plex: The expanded and complete set of TMTpro reagents for sample multiplexing. Journal of Proteome Research. 2021;**20**(5):2964-2972

[20] Everley PA, Krijgsveld J, Zetter BR, Gygi SP. Quantitative cancer proteomics: Stable isotope labeling with amino acids in cell culture (SILAC) as a tool for prostate cancer research. Molecular & Cellular Proteomics. 2004;**3**(7):729-735

[21] Walker SH, Papas BN, Comins DL, Muddiman DC. Interplay of permanent charge and hydrophobicity in the electrospray ionization of glycans. Analytical Chemistry. 2010;**82**(15):6636-6642

[22] Walker SH, Lilley LM, Enamorado MF, Comins DL, Muddiman DC. Hydrophobic derivatization of N-linked glycans for increased ion abundance in electrospray ionization mass spectrometry. Journal of the American Society for Mass Spectrometry. 2011;**22**(8):1309-1317

[23] King SR, Hecht ES, Muddiman DC. Demonstration of hydrazide tagging for O-glycans and a central composite design of experiments optimization using the INLIGHT™ reagent. Analytical and Bioanalytical Chemistry. 2018;**410**(5):1409-1415

[24] Mangrum JB, Mehta AY, Alabbas AB, Desai UR, Hawkridge AM. Comparative analysis of INLIGHT™-labeled enzymatically depolymerized heparin by reverse-phase chromatography and high-performance mass spectrometry. Analytical and Bioanalytical Chemistry. 2017;**409**(2):499-509

[25] Hecht ES, McCord JP, Muddiman DC. Definitive screening design optimization of mass spectrometry parameters for sensitive comparison of filter and solid phase extraction purified, INLIGHT plasma N-glycans. Analytical Chemistry. 2015;**87**(14):7305-7312

[26] Loziuk PL, Hecht ES, Muddiman DC. N-linked glycosite profiling and use of Skyline as a platform for characterization and relative quantification of glycans in differentiating xylem of *Populus trichocarpa*. Analytical and Bioanalytical Chemistry. 2017;**409**(2):487-497

[27] Kalmar JG, Butler KE, Baker ES, Muddiman DC. Enhanced protocol for quantitative N-linked glycomics analysis using Individuality Normalization

when Labeling with Isotopic Glycan Hydrazide Tags (INLIGHT)™. Analytical and Bioanalytical Chemistry. 2020;**412**(27):7569-7579

[28] Kalmar JG, Garrard KP, Muddiman DC. GlycoHunter: An open-source software for the detection and relative quantification of INLIGHT-labeled N-linked glycans. Journal of Proteome Research. 2021;**20**(4):1855-1863

[29] Hecht ES, Scholl EH, Walker SH, Taylor AD, Cliby WA, Motsinger-Reif AA, et al. Relative quantification and higher-order modeling of the plasma glycan cancer burden ratio in ovarian cancer case-control samples. Journal of Proteome Research. 2015;**14**(10):4394-4401

[30] Butler KE, Kalmar JG, Muddiman DC, Baker ES. Utilizing liquid chromatography, ion mobility spectrometry, and mass spectrometry to assess INLIGHT™ derivatized N-linked glycans in biological samples. Analytical and Bioanalytical Chemistry. 2022;**414**(1):623-637

[31] Hahne H, Neubert P, Kuhn K, Etienne C, Bomgarden R, Rogers JC, et al. Carbonyl-reactive tandem mass tags for the proteome-wide quantification of N-linked glycans. Analytical Chemistry. 2012;**84**(8):3716-3724

[32] Zhou S, Hu Y, Veillon L, Snovida SI, Rogers JC, Saba J, et al. Quantitative LC-MS/MS glycomic analysis of biological samples using AminoxyTMT. Analytical Chemistry. 2016;**88**(15):7515-7522

[33] Meitei NS, Apte A, Snovida SI, Rogers JC, Saba J. Automating mass spectrometry-based quantitative glycomics using aminoxy tandem mass tag reagents with SimGlycan. Journal of Proteomics. 2015;**127**(Pt A):211-222

[34] Yang S, Yuan W, Yang W, Zhou J, Harlan R, Edwards J, et al. Glycan analysis by isobaric aldehyde reactive tags and mass spectrometry. Analytical Chemistry. 2013;**85**(17):8188-8195

[35] Yang S, Wang M, Chen L, Yin B, Song G, Turko IV, et al. QUANTITY: An isobaric tag for quantitative glycomics. Scientific Reports. 2015;**5**:17585

[36] Xiang F, Ye H, Chen R, Fu Q, Li L. N,N-dimethyl leucines as novel isobaric tandem mass tags for quantitative proteomics and peptidomics. Analytical Chemistry. 2010;**82**(7):2817-2825

[37] Li M, Gu TJ, Lin X, Li L. DiLeuPMP: A multiplexed isobaric labeling method for quantitative analysis of O-glycans. Analytical Chemistry. 2021;**93**(28):9845-9852

[38] Li M, Zhong X, Feng Y, Li L. Novel isobaric tagging reagent enabled multiplex quantitative glycoproteomics via electron-transfer/higher-energy collisional dissociation (EThcD) mass spectrometry. Journal of the American Society for Mass Spectrometry. 2022;**33**(10):1874-1882

[39] Li M, Feng Y, Ma M, Kapur A, Patankar M, Li L. High-throughput quantitative glycomics enabled by 12-plex isobaric multiplex labeling reagents for carbonyl-containing compound (SUGAR) tags. Journal of Proteome Research. 2023;**22**(5):1557-1563

[40] Li M, Feng Y, Ma M, Li L. 12-Plex isobaric multiplex labeling reagents for carbonyl-containing compound (SUGAR) tag-enabled high-throughput quantitative glycomics. Methods in Molecular Biology. 2024;**2823**:155-172

[41] Ma M, Li M, Zhu Y, Zhao Y, Wu F, Wang Z, et al. 6-Plex. Analytical Chemistry. 2023;**95**(48):17637-17645

[42] Feng Y, Li M, Lin Y, Chen B, Li L. Multiplex quantitative glycomics enabled by periodate oxidation and triplex mass defect isobaric multiplex reagents for carbonyl-containing compound tags. Analytical Chemistry. 2019;**91**(18):11932-11937

[43] Shah P, Yang S, Sun S, Aiyetan P, Yarema KJ, Zhang H. Mass spectrometric analysis of sialylated glycans with use of solid-phase labeling of sialic acids. Analytical Chemistry. 2013;**85**(7):3606-3613

[44] Sun W, Zhang Q, Zhang X, Tran NH, Ziaur Rahman M, Chen Z, et al. Glycopeptide database search and de novo sequencing with PEAKS GlycanFinder enable highly sensitive glycoproteomics. Nature Communications. 2023;**14**(1):4046

Chapter 3

# Ionization Techniques for Mass Spectral Analysis

*Robert Owen Bussey III*

## Abstract

Mass spectral ionization techniques convert neutral molecules into charged species, enabling their detection and analysis by mass spectrometry. The techniques reviewed in this chapter are the following: electrospray ionization (ESI), atmospheric pressure chemical ionization (APCI), atmospheric pressure photoionization (APPI), and desorption electrospray ionization (DESI). All these techniques can be used in combination with chemical modifiers to ionize semi-volatile to non-volatile compounds and liquid chromatography and/or ion-mobility to separate compounds for mass spectral analysis. Explanations and examples of the ionization techniques will be given to show the significance to analytical chemistry. Additionally, the inclusion of distinct modifiers will be described as technique extensions that help with the ionization of compounds.

**Keywords:** mass spectral ionization, electrospray ionization (ESI), atmospheric pressure chemical ionization (APCI), atmospheric pressure photoionization (APPI), and desorption electrospray ionization (DESI)

## 1. Introduction

Mass spectrometers are analytical instruments with applications in the environmental, food forensic, natural products, and industrial sciences. A mass spectrometer always contains the following components: an ionization source to create ions from the compounds in a sample, a ion inlet entrance to the mass spectrometer, ion guides to create a path through the instrument, and a detector with data processing software to detect and analyze the ions. The main topic of this chapter is ionization techniques used to create ions from the samples in the mass spectrometer ionization source. These ions have a mass to charge ratio that is unique for each ionization technique and each compound based on the molecular weight and the amount of charge on the compound [1]. The fundamental principles behind the following ionization techniques will be reviewed in this chapter: electrospray ionization (ESI), atmospheric pressure chemical ionization (APCI), atmospheric pressure photoionization (APPI), and desorption electrospray ionization (DESI). Recent peer-reviewed studies that employ these ionization techniques will also be discussed to highlight their relevance in the alcohol, spice, and bioanalytical industries.

## 2. Ionization techniques explained

Compounds within samples must be ionized or charged for successful analysis in a mass spectrometer. The ionization techniques to create those charges such as ESI, APCI, APPI, and DESI are all classified as atmospheric pressure soft ionization techniques. These techniques are implemented at atmospheric pressure and not under vacuum like with electron impact ionization (EI). The techniques are considered 'soft' because they use low energy, which preserves the molecular ion and minimizes fragmentation [1]. The parent molecular ion is the intact chemical compound with an added charge without any fragmentation. A "hard" ionization technique such as electron impact (EI) causes extensive fragmentation of the compound resulting in limited detection of the intact parent compound [1].

Each ionization technique applies a charge to the compounds within sample to create ions in a unique way. These ions are transferred to the mass spectrometer with a vacuum pump with lower and lower vacuum as the ions near the analyzer [1]. The following sections will help explain the different ways samples are ionized in ESI, APCI, APPI, and DESI.

### 2.1 Electrospray ionization (ESI)

**Figure 1** provides a simplified schematic of electrospray ionization. With electrospray ionization, the sample is dissolved in a liquid mobile phase controlled by a high-pressure liquid chromatography (HPLC) system or through direct injection using a syringe containing the diluted sample. The ionization occurs at atmospheric pressure and an electric voltage between 2 and 5 kV is applied to the tip of the sprayer from which the sample comes out. This allows for a positive charge or a negative charge to be applied during different experiments or switching back and forth within the same experiment [1, 2]. Applying this charge at the sprayer tip causes charged droplets to break off from the liquid and form a shape called a Taylor Cone. In addition, heated

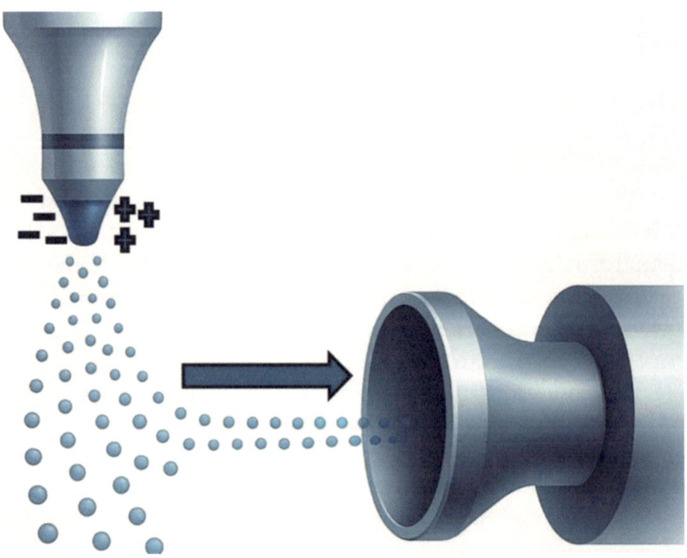

**Figure 1.**
*Simplified electrospray ionization representation.*

nitrogen gas guides the droplets out from the tip decreasing the droplets' volume as they travel [1]. As the droplet decreases in volume the droplet explodes leaving the charged ions to drift into the intake of the mass spectrometer driven by negative pressure [2].

During and after the ESI droplet explosion, there are several simple mechanisms that occur to transfer electrical charges to analytes. With ESI the charge creation occurs with polar and semipolar compounds and there are two charge states; a negative charge and a positive charge based on the electrical polarity. Acidic or basic functional groups in small and large molecules can lose or gain a positive proton based on the pH or additives contained in the droplets' solvent [2]. The proton can be added to a basic function group giving the molecule a positive charge or leave an acidic group giving the molecule a negative charge. This can be achieved by changing the pH with an organic acid or base like formic acid or ammonium hydroxide [2]. Fatty acids and lipids are some compounds that can be seen in negative mode. Amines, carbonyls, and polar phenolics are compounds that can be seen in positive mode.

In addition to the proton, other ions such as ammonium, formate, sodium, chloride, and other salts can complex with acidic/basic groups or with other polar groups to form a charge [2]. In-source fragmentation such as loss of water can occur even though ESI is a soft ionization technique. There are some additional mechanisms for charge transfer but these are out of the scope of this book chapter.

### 2.2 Atmospheric pressure chemical ionization (APCI)

Atmospheric pressure chemical ionization, as the name suggests, runs at atmospheric pressure, and as in ESI the samples are dissolved in a liquid mobile phase controlled by a high-pressure liquid chromatography (HPLC) system or through direct injection using a syringe containing the diluted sample [1, 3]. The difference from ESI is that the samples are injected into the ionization source using a nebulizer

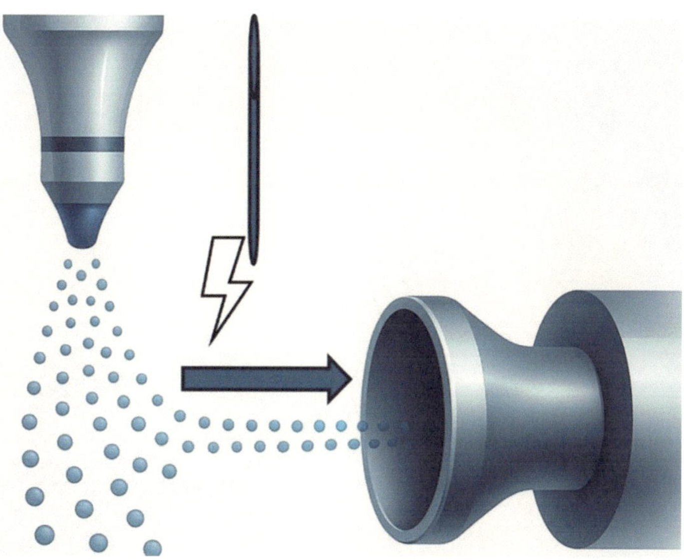

**Figure 2.**
*Atmospheric pressure chemical ionization representation which uses a corona discharge to ionize the gaseous compounds.*

with heated nitrogen gas that creates a droplet fog. **Figure 2** shows a simple representation of how the droplets are then charged using a corona discharge from a needle in contact with the droplet fog. The high voltage from the needle causes the corona discharge which is an ionized gaseous plasma [4]. The application of the hot nitrogen gas makes the droplets smaller to ionize the solvent and compound peaks better. The corona discharge allows for proton addition/adduct formation to create a positive ion, or proton extraction/adduct formation to create a negative ion [1]. The evaporated mobile phase also acts as a charge carrier thus ionizing the compounds within the droplets rather than the droplets exploding from excess charge. In addition, side reactions with nitrogen and oxygen radicals can occur with increased corona discharge current and desolvation gas temperature [5]. These can create artifacts that differentiate from the protonated or deprotonated species. The extra descriptive data can also provide additional identification options for pure standards in matrix. Just like with ESI, the droplets would be charged and then pulled into the mass spectrometer for analysis. Both polar and relatively nonpolar compounds are easier to ionize with APCI than with ESI, but just like with ESI, some in-source fragmentation such as water loss can occur even though it is a soft ionization technique. Many of the ionization techniques can be used as complimentary ionizing methods [1].

### 2.3 Atmospheric pressure photoionization (APPI)

Atmospheric pressure photoionization is like ESI and APCI in that samples are dissolved in a liquid mobile phase controlled by a high-pressure liquid chromatography (HPLC) system or through direct injection using a syringe containing the diluted sample. Like APCI the liquid samples goes through a heated nebulizer with heated nitrogen gas that creates a droplet fog, but the only difference is the lack of a corona discharge [1]. Instead, photons from an ultraviolet source ionize compounds in the droplets [6]. The advantage of this technique is that it is more selective at ionizing the compounds from the sample rather than the solvent and the atmospheric gas which gives less artifacts and background noise compared to APCI [1]. On the other hand,

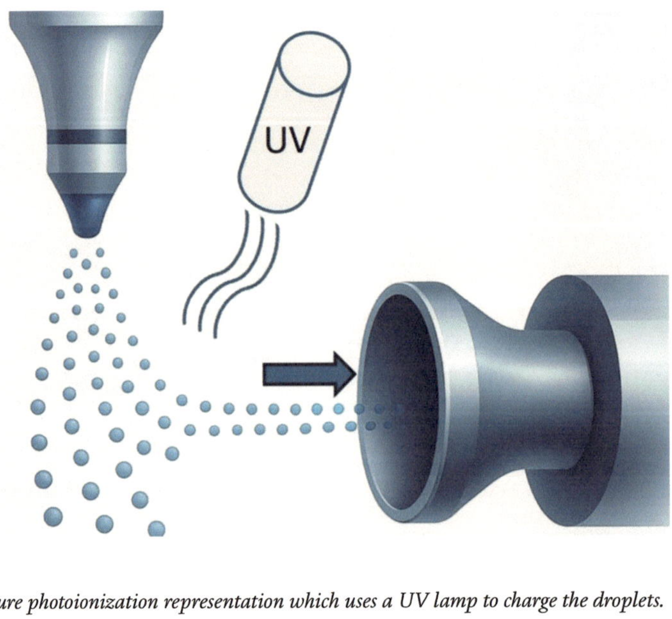


**Figure 3.**
*Atmospheric pressure photoionization representation which uses a UV lamp to charge the droplets.*

there can be lower ionization efficiency of the sample compounds due to the absorption of a portion of the photon energies by the gaseous solvent (**Figure 3**).

The photon ionization of the sample causes a radical cation formation (charge exchange) or a protonated ion species (like ESI and APCI) for positive mode. For negative mode, ionization occurs through radical anion formation (charge exchange) or proton extraction. Radical cation formation usually occurs in solvents that have low proton affinity in the gas phase such as water, chloroform, and cyclohexane while proton transfer occurs with solvents higher in proton affinity in the gas phase such as methanol and acetonitrile [1, 6]. When there is a mix of these solvents, there may be a mix of charged species. Also, there are caveats to this ionization scheme relative to ionization affinities of the compound versus the solvent [6, 7]. In addition, dopants can be used for both positive and negative ion mode but are more important in negative mode ionization accepting electrons and creating radical anions through electron capture. These dopants such as toluene, acetone, and chloroform can then transfer their electrons to the compounds in the sample if the compounds have high electron affinity [1, 6]. Oxygen atoms can also be ionized into radical anions through charge exchange and then deprotonate compounds with high gas phase acidity [1, 6]. As can be seen through this section, many different things can affect the ionization in APPI, but unlike ESI and APCI more non-polar compounds can be seen through APPI [1, 7]. Using APPI alongside ESI and APCI expands the range of polarities for an experiment if required.

### 2.4 Desorption electrospray ionization

**Figure 4** shows the basic principle of desorption electrospray ionization (DESI) which is similar to electrospray ionization. The ionization occurs at atmospheric pressure and an electric current between 2 and 5 kV is applied to the tip of the

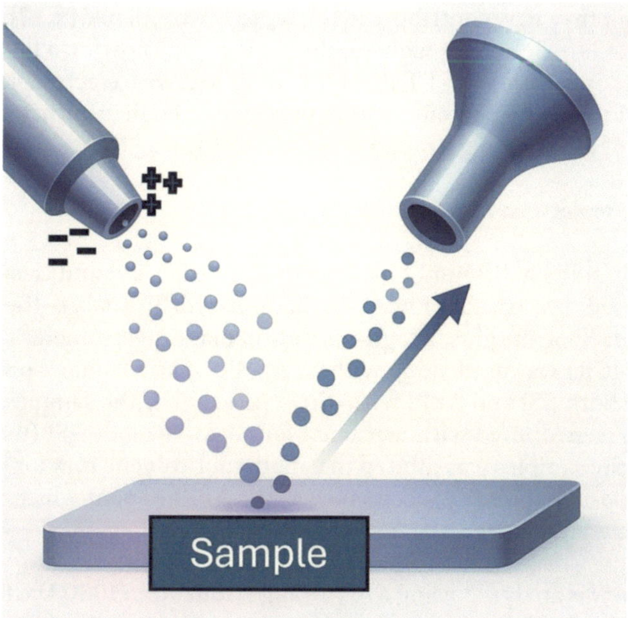


**Figure 4.**
*Desorption electrospray ionization representation.*

sprayer from which solvent only comes out [1, 8]. This allows for a positive charge or a negative charge to be applied to the solvents during different experiments or switching back and forth within the same experiment [1]. As in ESI, charged droplets break off from the tip and are propelled with heated nitrogen gas. The difference as seen in **Figure 4** is that the charged solvent particles are sprayed on the surface of the sample which is on a flat surface [1, 8]. The charged solvent ions desorb and ionize compounds from the sample which then bounce off the surface and into the inlet of the mass spectrometer. The solvent can be optimized to ionize specific compounds in a sample by changing the solvent or the additives as in ESI. The ionization mechanism is like ESI in which there are pronated and deprotonated species, single/multiple charges based on number of functional groups, adduct formations with salt ions/additives, and possible in-source fragmentations [1].

## 3. Literature review of multiple ionization techniques in action

### 3.1 Electrospray ionization and atmospheric pressure photoionization

#### *3.1.1 Background*

Gin is famous globally for its unique taste with a projected eight hundred million liters of gin sold in 2021. It is made by redistilling grain-based alcohol in the presence of botanicals with each manufacturer having their unique blends. The main criteria and flavor component is Juniper berries with other secondary botanicals including coriander, citrus peels, and other herbs [9–11]. The mixture of botanicals used creates a complex chemical profile including volatile and non-volatile compounds. Gas chromatography mass spectrometry (GC-MS) has been used to analyze many of the volatile components in gin, but nonvolatile constituents have not been well characterized even though they may contribute to the sensory experience [9, 12]. This study assessed if direct infusion using high-resolution Fourier Transform Ion Cyclotron Resonance Mass Spectrometry (FT-ICR MS) combined with electrospray ionization and atmospheric pressure photoionization provided enough information for a chemical fingerprint of gin covering a broad range of volatilities [9].

#### *3.1.2 Methods, samples, and instrumentation*

In this study, sixteen different gin alcohols were analyzed and classified as either grain-based, rye-based, or wine distillate-based. To analyze the chemical compounds in each of the gins, a high-resolution mass spectrometer (12-T Bruker SolariX XR FT-ICR) was used along with an Apollo-II atmospheric pressure source that can utilize both ESI and APPI with direct infusion of the samples. Small amounts of gin were diluted with methanol for negative mode ESI (using a negative polarity), and the samples was diluted in a methanol-toluene mix for positive mode APPI (using a positive polarity). Toluene was the APPI dopant which increased ionization efficiency. The specific ionization settings are beyond the scope of this chapter.

Each sample was analyzed using a mass range from 90 to 1000 Da. for compound detection, and the data was processed with Bruker DataAnalytics 5.0 software. The output of the analysis yielded accurate masses that were searched against CompoundCrawler to suggest structures with specific confidence criteria.

### *3.1.3 Results*

This study distinguished between compounds from direct infusion using high resolution mass spectrometry to differentiate the individual chemical compounds. As discussed in the ionization overview section, ESI ionizes the polar, heteroaromatic compounds as can be seen with the oxygen-containing compounds such as the selected compounds in **Figure 5**. Many compounds (1300–5200 unique features) were detected but only a fraction is discussed in the study. It is important to note that some of the features seen may be repeat signals of the same compound since multiple adduct peaks of the same compound can occur. The most abundant peak detected was palmitic acid and other long chain fatty acids which are to be expected as common from botanical sources. The high abundance of gingerol was unexpected in the study since the gin's major component is juniper berries and gingerol has not previously been linked to juniper. In addition, terpenoids which contribute to the flavor such as juvabione, linalool, and sclareol were also detected and are pictured in **Figure 5**. Other phenolic compounds such as syringaldehyde, sinapaldehyde were present at high intensities in one gin sample because it was aged in an oak barrel. The gin extracted in the oak-derived phenolics would have seen even more phenolics if the barrel was charred.

This study used APPI as a complimentary ionization method to visualize the more nonpolar compounds which ESI cannot ionize. This covered a wider range of polarity specifically the compounds in **Figure 6**. These compounds (terpenes and esters) based on the non-polar characteristics would normally be analyzed by gas chromatography mass spectrometry but are seen using APPI. For compounds that contain a heteroatom, the most abundant compound peaks typically include the radical cation and/or the protonated species.

This study detected around 1600 features per gin with a greater than 3100 features in the oak-barrel-aged gin. The data not shown here includes tentative identifications of the following chemical classes: monoterpenes, sesquiterpenes, diterpenes, phenolic compounds, alcohols, ketones, aldehydes, and esters. These compounds may

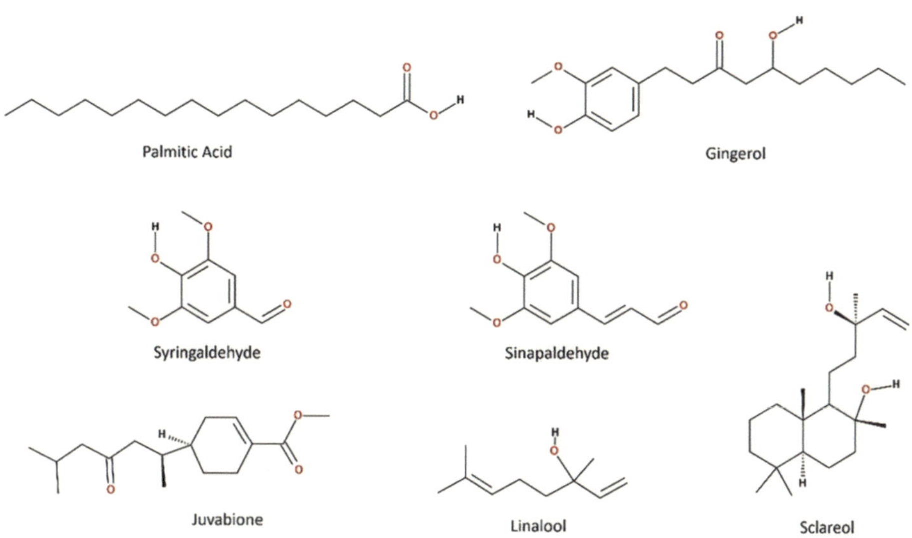


**Figure 5.**
*Selected compounds seen in gin using negative mode ESI.*

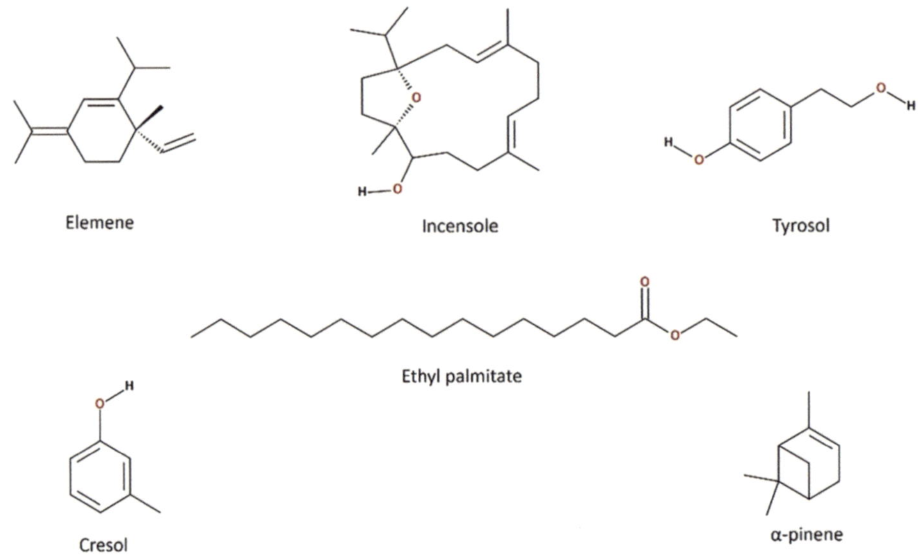


**Figure 6.**
*Selected compounds seen in gin using positive mode APPI.*

contribute more to the sensory experience based on their high volatility. Future work could be done to use a chromatography technique to separate the compounds because the monoisotopic masses cannot differentiate between structural isomers (compounds with the same formula but different structures). The statistics performed for this work are beyond this chapter but were used to find similarities and differences in spectral features. The findings showed that using complementary techniques one could see additional compounds not previously reported. Only six of the compounds found in this study were previously reported in gin. It is important to report that of the many features seen, 135 compounds were identified according to their identification criteria reported in the study (not shown here). 122 of the 135 compounds had not been reported in gin before.

### *3.1.4 Study conclusions*

This study used direct-infusion high-resolution FT-ICR mass spectrometry to analyze gin which is an extremely popular spirit worldwide. Gin's signature flavor comes from a mix of natural botanicals. The objective of this study was to explore the chemical fingerprint complexity.

Sixteen commercial gins were examined using a combination of ESI and APPI techniques. This revealed detailed chemical fingerprints for each sample, identifying 135 different volatile and non-volatile compounds. Of the identified compounds, 122 had never been reported in gin before such as gingerol. The application of ESI was significant by expanding to the nonvolatile compounds which had previously been unexplored.

While this method is an efficient and detailed way to profile the chemical composition of gin, it does have limitations. It cannot distinguish between structural isomers based on mass alone. To resolve those cases, additional techniques like GC-MS, LC-MS, or tandem mass spectrometry could be applied.

## 3.2 Electrospray ionization, atmospheric pressure chemical ionization, and atmospheric pressure photoionization

### *3.2.1 Background*

Color of a food product is a major criterion in determining acceptance of that product by a consumer. To increase consumer acceptance, colored dyes may be added to food to re-color the food to its original color if it has changed during food processing or manufacturing [13]. The use of natural dyes is favored in many markets based on regulation, but synthetic dyes in food products is common since they can be lower cost and potentially more stable. Synthetic dyes used in spices have raised significant health concerns due to their potential teratogenic and genotoxic properties [13]. Dyes such as Sudan I–IV, Sudan Orange G, Sudan Red 7B, Para Red, and Rhodamine B are banned in food products by regulatory bodies like the European Union due to their potential toxicity [14–16]. Despite this, these dyes are still illicitly used to enhance the visual appeal of spices like turmeric, curry, and chili.

Previous analytical methods, including liquid chromatography ultra-violet (LC-UV) and liquid chromatography tandem mass spectrometry (LC-MS/MS), have been employed to detect these dyes [13]. However, most studies have focused on electrospray ionization, with limited exploration of atmospheric pressure chemical ionization and atmospheric pressure photoionization, which may offer advantages in terms of matrix effects and ionization efficiency. This study compares ESI, APCI, and APPI to identify advantages of each when analyzing the following synthetic dyes Sudan I–IV, Sudan Orange G, Sudan Red 7B, Para Red, and Rhodamine B [13].

### *3.2.2 Methods, samples, and instrumentation*

Eight dyes were studied: Sudan I–IV, Sudan Orange G, Sudan Red 7B, Para Red, and Rhodamine B. Stock solutions were prepared in acetonitrile, DMSO, or acetone depending on solubility. Matrix match analyses were done to make sure to minimize the matrix effects in turmeric, curry, and chile.

The detection of the dyes used an Ultra High Pressure Liquid Chromatography system (UHPLC) system coupled to a triple quadrupole mass spectrometer (TSQ Quantum Ultra AM). An Accucore C18 (100 mm × 2.1 mm, 2.6 μm) was used to separate the dyes from the matrix interferences if any. The gradient included formic acid/ammonium formate buffers with acetonitrile (pH 3.75) starting from high aqueous to high organic solvent. The specifics of the gradient are beyond the scope of this chapter.

Three ionization sources were assessed: ESI, APCI, and APPI (with dopants like acetone, toluene, tetrahydrofuran, anisole, and chlorobenzene). Refer to the study for specifics of the settings of the ionization sources. Thirty-six spice samples of turmeric, curry, chili from various countries were evaluated for the amounts of the eight dyes in **Figure 7**. The mass spectral data was acquired in both full scan and multiple reaction monitoring mode (MRM) to quantitate the levels of the dyes. In addition, fragmentation studies were done on a linear ion trap-Orbitrap Velos using the APCI source in both positive and negative modes to increase compound detection.

The extraction consisted of solubilizing five grams of sample in 5 mL of acetonitrile, vortexing, sonicating, centrifuging, and filtering the final sample prior to analysis. This was repeated for all the samples for analysis. In addition, each of the dyes was added to an example sample of turmeric, curry, and chile to estimate

| Compound | R1 | R2 | R3 | R4 | R5 | Chemical Formula | Monoisotopic Mass |
|---|---|---|---|---|---|---|---|
| Sudan I | -OH | -H | -H | -H | -H | $C_{16}H_{12}N_2O$ | 248.0944 |
| Sudan II | -OH | -H | $-CH_3$ | -H | $-CH_3$ | $C_{18}H_{16}N_2O$ | 276.1257 |
| Sudan III | -OH | -H | -N=N-Ph | -H | -H | $C_{22}H_{16}N_4O$ | 352.1318 |
| Sudan IV | -OH | $-CH_3$ | -N=N-Ph-*o*$CH_3$ | -H | -H | $C_{24}H_{20}N_4O$ | 380.1631 |
| Sudan Red 7B | -NH-$CH_2CH_3$ | -H | -N=N-Ph | -H | -H | $C_{24}H_{21}N_5$ | 379.1791 |
| Para Red | -OH | -H | -N-$O_2$ | -H | -H | $C_{16}H_{11}N_3O_3$ | 293.0794 |

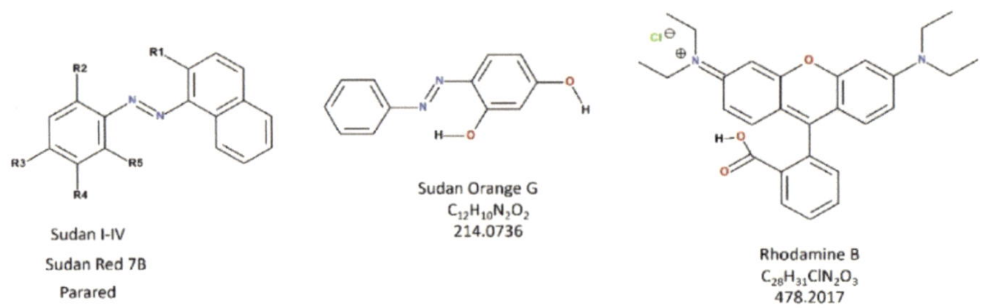


**Figure 7.**
*Structures of Sudan I–IV, Sudan Orange G, Sudan Red 7B, Para Red, and Rhodamine B. Functional groups for related dyes are used to help simplify the representation.*

extraction efficiency along with limits of quantitation (LOQ) and limits of detection (LOD) for each dye.

### *3.2.3 Results*

The data was gathered with a baseline assumption that the ESI would have more matrix interferences and background noises compared to the APCI and APPI based on past works [13]. ESI, APCI and APPI were all evaluated in both positive and negative ionization modes in optimal conditions for each. **Table 1** shows different ionization sources with each modes' most abundant ions. The Sudan dyes all showed the protonated ion $[M + H]^+$ as the most intense signal for ESI, APCI, and APPI. For all ionization sources, Rhodamine B showed the molecular ion with no proton lost or gained $[M]^+$ because it is already charged as a quaternary ammonium salt. Para Red had the best signal in negative-ion mode for all ionization sources. ESI did not have any adduct formations nor in-source fragmentation. APCI had in-source fragmentation for Sudan II and Sudan Red 7B with less than 30% relative abundance while still in the presence of the main protonated ion.

| Compound | ESI | | APCI | | APPI | |
|---|---|---|---|---|---|---|
| | m/z (Rel. Ab. %) | Ion Assignment | m/z (Rel. Ab. %) | Ion Assignment | m/z (Rel. Ab. %) | Ion Assignment |
| Sudan I | 249.1 (100) | $[M + H]^+$ | 249.1 (100) | $[M + H]^+$ | 249.1 (100) | $[M + H]^+$ |
| | | | | | 156.1 (16) | $[M + H\text{-}C_6H_5O]^+$ |
| Sudan II | 277.1 (100) | $[M + H]^+$ | 277.1 (100) | $[M + H]^+$ | 277.1 (100) | $[M + H]^+$ |
| | | | 156.1 (10) | $[M + H\text{-}C_8H_{11}N]^+$ | 156.1 (21) | $[M + H\text{-}C_8H_{11}N]^+$ |
| | | | 120.1 (12) | $[M + H\text{-}C_{10}H_7NO]^+$ | 120.1 (25) | $[M + H\text{-}C_{10}H_7NO]^+$ |
| Sudan III | 353.2 (100) | $[M + H]^+$ | 353.2 (100) | $[M + H]^+$ | 353.2 (100) | $[M + H]^+$ |
| Sudan IV | 381.2 (100) | $[M + H]^+$ | 381.2 (100) | $[M + H]^+$ | 381.2 (100) | $[M + H]^+$ |
| | | | | | 224.2 (14) | $[M + H\text{-}C_{10}H_7NO]^+$ |
| Sudan Orange | 215.1 (100) | $[M + H]^+$ | 215.1 (100) | $[M + H]^+$ | 215.1 (100) | $[M + H]^+$ |
| Sudan Red 7B | 380.2 (100) | $[M + H]^+$ | 380.2 (100) | $[M + H]^+$ | 380.2 (100) | $[M + H]^+$ |
| | | | 183.1 (36) | $[M + H\text{-}C_{12}H_{11}N_3]^+$ | 183.1 (36) | $[M + H\text{-}C_{12}H_{11}N_3]^+$ |
| Rhodamine B | 443.2 (100) | $[M]^+$ | 443.2 (100) | $[M]^+$ | 443.2 (100) | $[M]^+$ |
| Para Red | 292.1 (100) | $[M\text{-}H]^-$ | 292.1 (100) | $[M\text{-}H]^-$ | 292.1 (100) | $[M\text{-}H]^-$ |

**Table 1.**
*Results from the different ionization modes for eight synthetic dyes.*

For APPI a dopant was not needed to ionize the compounds in positive mode, but a dopant was needed in negative mode to help with the ionization. As mentioned in a previous section APPI requires a dopant in negative ionization mode. The dopant to produce the largest signal in APPI was the chlorobenzene. All protonated species and no radical cations were formed. Sudan I, I, IV, Sudan 7B had in source fragmentation in the protonated forms as seen in **Table 1**. The fragmentation patterns were confirmed with the Oribtrap to confirm the structural losses. Additional fragmentation experiments were performed on a tandem quadrupole mass spectrometer to create appropriate methods for quantitation using the fragmentation patterns to increase sensitivity and specificity.

To evaluate matrix effects, the eight dyes were spiked into a spice powder extract, shaken for 12 hours, and then extracted with acetonitrile which proved to be the best solvent to give the least matrix effect. Both low and high concentrations were spiked into separate samples to help further confirm matrix effects. None of the analyzed samples contained the banned dyes. Based on matrix effects (ion suppression or enhancement) and ease of use, APCI was chosen as the best ionization source for the eight banned dyes.

#### *3.2.4 Conclusions*

This study compared three ionization techniques—ESI, APCI, and APPI—to determine which one works best for detecting eight banned azo and non-azo dyes in spices using LC-MS/MS. The researchers also looked at how well a quick and simple extraction method worked with each technique, considering how compounds from spices like turmeric, curry, and chile might interfere with detection. After evaluating 36 spice samples, they found that the UHPLC-APCI-MS/MS method was the most effective for reliably identifying these illegal dyes.

### 3.3 Desorption electrospray ionization

#### *3.3.1 Background*

Microbes have played a key role in industry for decades in areas such as alcohol production through fermentation and antibiotic production. With the rise of genetic engineering tools like recombinant DNA and CRISPR, scientists can now genetically alter microbes to make all kinds of valuable chemicals [17, 18]. A common use is to boost production of free fatty acids (FFAs) in organisms like *Escherichia coli* (*E. coli*) by modifying their metabolism. These engineered strains could potentially serve as an additional source for biofuels [19, 20]. The genetic modification process is a complicated process because over production of a group of analytes could cause problems in other microbial pathways. Analytical testing is needed to confirm analytes of interest and monitor any unintended production pathways.

Mass spectrometry coupled with a chromatography technique is a perfect tool but depending on the sample preparation and time for analysis, the throughput might not be high enough. Desorption Electrospray mass spectrometry (DESI-MS), allows for fast, spatially resolved analysis under ambient conditions [21]. It can speed up secondary metabolite characterization and reduces sample prep by analyzing bacteria or other solid samples on a surface [21].

In this study, the researchers created a DESI-MS "Fast-Pass" workflow to decrease analysis time by putting different samples along the length of a slide to collect data as the slide moved length wise. The method was set up in a way to identify the secondary metabolites created by the engineered microbes 10X faster than other methods [21].

### *3.3.2 Methods, samples, and instrumentation*

Three *E. coli* strains were used: TY05 (produces mostly C12:0 fatty acids), NHL17 (produces C8:0), and TY04 (a control strain). Cultures were started from freezer stocks, grown overnight, and then expanded into larger volumes. Once the cultures reached a certain stage of growth, they were induced with IPTG (Isopropyl β-D-1-thiogalactopyranoside) to trigger fatty acid production and incubated for 24 hours. Afterward, the cultures were flash-frozen for later analysis by DESI-MS.

To evaluate the Fast-Pass method, standard mixtures of peptides, sugars, fatty acids, and small molecules were prepared in methanol/water. These were spotted onto a PTFE (Polytetrafluoroethylene) grid and left to dry. The PTFE allows the sample to dry with little to no chemical interaction with the slide. Each row of dots contained eleven spots separated over a 50 mm length. Sugars were mixed with sodium acetate to help with ionization. Peptides, sugars (maltose plus others), labeled lipids, and small molecules (caffeine and reserpine) were analyzed in positive ion mode, while fatty acids were analyzed in negative mode.

For the bacterial samples, frozen cultures were thawed and processed using a liquid-liquid extraction method meant to extract the lipophilic (fatty) class of compounds. The bacterial amount was confirmed and then the liquid-liquid partitioning was performed. This involved crashing proteins with cold methanol, extracting lipids using a mix of (methyl tert-butyl ether) MTBE, methanol, and water, and adding internal standards for quality control. The extracted lipids in the MTBE layer were dried, re-dissolved in 90:10 methanol/chloroform, and spotted onto a PTFE-coated slide in a randomized layout. A blank sample of growth media was also included for background correction.

The mass spectrometer used in this experiment was a Waters Synapt G2-S high-definition mass spectrometer with ion mobility with a DESI-MS source (prototype at this time). Currently the DESI-MS source is commercialized and is not a prototype. The ion mobility separation is a form of separation that on the millisecond scale occurs after the sample enters the mass spectrometer [22]. The nitrogen gas for DESI was heated to about 150°C and voltage across the cone was 5 kV. This is important because these two parameters allow for the ionization of the 90:10 acetonitrile:water solvent so that it could ionize the sample on the PTFE slide. Just like with ESI, formic acid is included for positive-ion mode and an ammonium salt is included for negative-ion mode. The angle of the spray related to the sprayer and the mass spectrometer inlet was optimized for best signal.

### *3.3.3 Results*

Traditional DESI-MS imaging takes a long time because it scans samples row by row to create detailed images. The setup of the samples on the slide in a linear one-dimensional fashion eliminated the complex movements that would be required if the sprayer had to move in two-dimensions. The researchers scanned once across all the samples with each line a replicate of each sample. This balanced the need for fast measurement with the information acquired during the scan. The results are like a chromatogram where signal (y-axis) is compared to time (x-axis), but here the x-axis is distance in mm of the plate. Each sample's position is known and easily compared.

To assess the method, the study used mixtures of known compounds standards. Each compound showed up clearly in the chromatogram, with little to no overlap between samples. Differences in signal strength were mostly due to how well each compound ionized. The way compounds dried on the spots also affected their distribution with peptides tending to concentrate on the edges, sugars in the center, and others spread evenly. The differences in compound distribution could be seen on the mass spectral readout since the DESI was based off the reading across a dried droplet. Despite these differences, the signals were consistent and reproducible across samples. Also, most signals showed $[M + H]^+$ and $[M\text{-}H]^-$ for the observed signals.

The speed at which the sample stage moved (raster rate) affected how much signal was collected. Faster movement meant less time per sample and less signal intensity. The team tested different speeds and found that 125 μm/s gave the best balance between speed and signal quality. This rate allowed them to analyze each sample in about 40 seconds.

Three *E. coli* strains were used: TY05 (produces mostly C12:0 fatty acids or dodecanoic acid), NHL17 (produces C8:0 or octanoic acid), and TY04 (a control strain). The Fast-Pass method successfully identified the expected fatty acids in each strain (**Figure 8**) and collected non-targeted data for any other analyte signals in each sample. It also detected side products like C12:1 (dodecenoic acid) and C14:1 with double bonds, which are known byproducts of the genetic modifications. C14:0 fatty acids (tetradecanoic acid) were also unexpectedly seen in some strains showing that a full scan data collection in addition to targeted scans can be used to create a holistic picture of the compounds present in the samples.

The method produced consistent results across different days and sample replicates. Internal standards (labeled lipids) were used to check for variability, and the results showed less than 10% variation within replicates. This confirmed that Fast-Pass was dependable for screening engineered microbes. A statistical analysis method

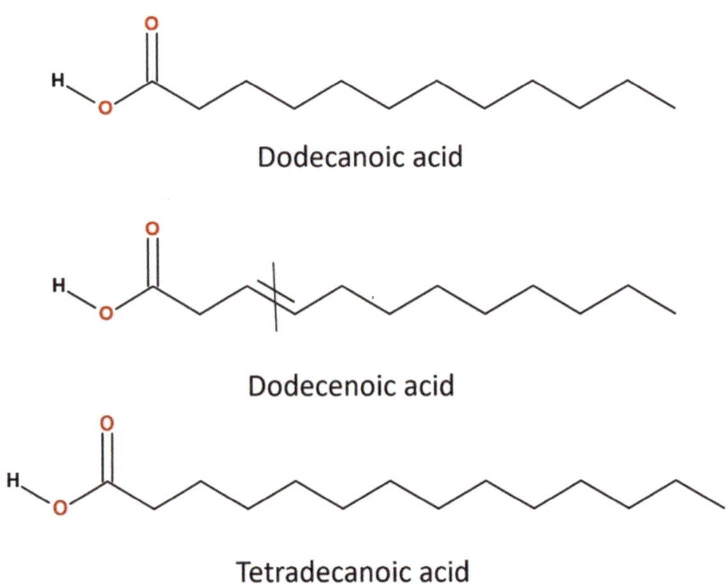


**Figure 8.**
*Examples of fatty acid products produced by genetically modified* E. coli.

called Principal Component Analysis (PCA) to group the samples based on their metabolic profiles. Each strain showed distinct PCA groupings and the engineered strains had more significant changes compared to the control. Other stats helped identify which compounds were up- or down-regulated due to genetic edits. TY05 showed more changes than NHL17, suggesting its modifications had a bigger impact on metabolism. This data showed that DESI-MS creates a complex data profile that gave enough information for a metabolomics analysis of the secondary metabolites.

#### *3.3.4 Conclusions*

The Fast-Pass DESI-MS method offered a fast and reliable way to screen genetically modified microbes. It simplified the process by scanning a line of samples and avoids the need for complex image analysis. The method works across diverse types of compounds and consistently produced detailed chemical profiles in under a minute per sample.

When applied to engineered *E. coli*, Fast-Pass accurately identified fatty acid production and broader metabolic changes. It significantly sped up the analysis process, helping researchers keep pace with rapid developments in synthetic biology.

As seen in the study Fast-Pass is not limited to fatty acids. It can be used in many biological systems and integrated into existing workflows. It also complements deeper analytical methods like LC-IM-MS, making it a valuable tool for both quick screening and detailed investigations.

## 4. Conclusion

This chapter explored how widely used atmospheric pressure ionization technique, ESI, APCI, APPI, and DESI could be applied to modern mass spectrometry. Each method has its strengths depending on the type of compounds being analyzed, the sample matrix, and the goals of the analysis.

Through real-world examples, this chapter explores how these techniques are used in practice such as identifying complex botanical compounds in gin, to detecting banned dyes in spices, and rapidly screening genetically engineered microbes. These case studies show that no single ionization method fits all situations. Instead, combining techniques often gives a more complete picture, especially when dealing with complex or diverse samples.

The chapter also looked at how newer workflows like Fast-Pass DESI-MS can dramatically speed up analysis without sacrificing data quality. This kind of innovation is helping analytical chemistry keep pace with the fast-moving fields of synthetic biology, food safety, and environmental science.

In short, understanding how and when to use each ionization technique is key to getting the most out of mass spectrometry. As instrumentation continues to evolve, so will the ways we apply these tools to solve real-world problems.

## Conflict of interest

The author declares no conflict of interest.

## Author details

Robert Owen Bussey III
Reynolds American Inc., Winston Salem, North Carolina, United States of America

*Address all correspondence to: busseyr@rjrt.com

## References

[1] De Hoffmann E, Stroobant V. Mass Spectrometry: Principles and Applications. 3rd ed. Chichester: John Wiley and Sons; 2007

[2] Cech N, Enke C. Practical implications of some recent studies in electrospray ionization fundamentals. Mass Spectrometry Reviews. 2001;**20**:362-387. DOI: 10.1002/mas.10008

[3] French JB, Thomson BA, Davidson WR, Reid NM, Buckley JA. Atmospheric pressure chemical ionization mass spectrometry. In: Karasek FW, Hutzinger O, Safe S, editors. Mass Spectrometry in Environmental Sciences. Boston, MA: Springer; 1985. DOI: 10.1007/978-1-4613-2361-7_6

[4] Lloyd JR, Hess S. A corona discharge initiated electrochemical electrospray ionization technique. Journal of the American Society for Mass Spectrometry. 2009;**20**:1988-1996. DOI: 10.1016/j.jasms.2009.07.021

[5] Samarasinghe I, Pavlov J, Attygalle AB. Unexpected artifact formation in mass spectrometric analysis of aniline under atmospheric-pressure chemical ionization. Journal of the American Society for Mass Spectrometry. 2025;**36**:463-472. DOI: 10.1021/jasms.4c00286

[6] Kauppila TJ, Kotiaho T, Kostiainen R, Bruins AP. Negative ion-atmospheric pressure photoionization-mass spectrometry. Journal of the American Society for Mass Spectrometry. 2004;**15**:203-211. DOI: 10.1016/j.jasms.2003.10.012

[7] Short LC, Cai SS, Syage JA. APPI-MS: Effects of mobile phases and VUV lamps on the detection of PAH compounds. Journal of the American Society for Mass Spectrometry. 2007;**18**:589-599. DOI: 10.1016/j.jasms.2006.11.004

[8] Towers MW, Karancsi T, Jones EA, Pringle SD, Claude E. Optimised desorption electrospray ionisation mass spectrometry imaging (DESI-MSI) for the analysis of proteins/peptides directly from tissue sections on a travelling wave ion mobility Q-ToF. Journal of the American Society for Mass Spectrometry. 2018;**29**:2456-2466. DOI: 10.1007/s13361-018-2049-0

[9] Dou Y, Mäkinen M, Jänis J. Analysis of volatile and nonvolatile constituents in gin by direct-infusion ultrahigh-resolution ESI/APPI FT-ICR mass spectrometry. Journal of Agricultural and Food Chemistry. 2023;**71**:7082-7089. DOI: 10.1021/acs.jafc.3c00707

[10] Gin Bk. In: Russell I, Stewart GG, Kellershohn J, editors. Whisky and Other Spirits. 3rd ed. Oxfird, UK: Elsevier; 2021. 576 p. DOI: 10.1016/C2019-0-03286-4

[11] Vichi S, Riu-Aumatell M, Mora-Pons M, Guadayol JM, Buxaderas S, Lopez-Tamames E. HS-SPME coupled to GC/MS for quality control of *Juniperus communis L.* berries used for gin aromatization. Food Chemistry. 2007;**105**:1748-1754. DOI: 10.1016/j.foodchem.2007.03.026

[12] Buck N, Goblirsch T, Beauchamp J, Ortner E. Key aroma compounds in two Bavarian gins. Applied Sciences. 2020;**10**:7269. DOI: 10.3390/app10207269

[13] Arrizabalaga-Larrañaga A, Epigmenio-Chamú S, Santos FJ, Moyano E. Determination of banned

dyes in red spices by ultra-high-performance liquid chromatography-atmospheric pressure ionization-tandem mass spectrometry. Analytica Chimica Acta. 2021;**1164**:338519. DOI: 10.1016/j.aca.2021.338519

[14] Regulation E. No 1333/2008 of the European Parliament and of the council of 16 December 2008 on food additives. Official Journal of the European Union. 2008;**354**:16-33

[15] E Commission. Commission decision of 23 May 2005 on emergency measures regarding chilli, chilli products, curcuma and palm oil. Official Journal of the European Union. Communities: Legis. 2005;**402**:34-35

[16] Mpountoukas P, Pantazaki A, Kostareli E, Christodoulou P, Kareli D, Poliliou S, et al. Cytogenetic evaluation and DNA interaction studies of the food colorants amaranth, erythrosine and tartrazine. Food and Chemical Toxicology. 2010;**48**:2934-2944. DOI: 10.1016/j.fct.2010.07.030

[17] Jullesson D, David F, Pfleger B, Nielsen J. Impact of synthetic biology and metabolic engineering on industrial production of fine chemicals. Biotechnology Advancements. 2015;**33**:1395-1402. DOI: 10.1016/j.biotechadv.2015.02.011

[18] Jiang F, Doudna JA. CRISPR-Cas9 structures and mechanisms. Annual Review of Biophysics. 2017;**46**:505-529. DOI: 10.1146/annurev-biophys-062215-010822

[19] Marella ER, Holkenbrink C, Siewers V, Borodina I. Engineering microbial fatty acid metabolism for biofuels and biochemicals. Current Opinion in Biotechnology. 2018;**50**:39-46. DOI: 10.1016/j.copbio.2017.10.002

[20] Lennen RM, Pfleger BF. Engineering *Escherichia Coli* to synthesize free fatty acids. Trends in Biotechnology. 2012;**30**(12):659-667. DOI: 10.1016/j.tibtech.2012.09.006

[21] Shepard HS, May JC, Zuniga BE, Abraham JP, Pfleger BF, Young JD, et al. A fast-pass, desorption electrospray ionization mass spectrometry strategy for untargeted metabolic phenotyping. Journal of the American Society for Mass Spectrometry. 2025;**36**:400-408. DOI: 10.1021/jasms.4c00459

[22] Owen Bussey R. Perspective chapter: Advantages of ion mobility coupled with HPLC/UPLC. In: Analytical Liquid Chromatography - New Perspectives. London, UK: IntechOpen; 2022. DOI: 10.5772/intechopen.102380

Chapter 4

# Mass Spectrometry Applications in Forensic Sciences

*Dilek Salkım İşlek and Eda Kiriş*

## Abstract

Forensic science is a necessary part of contemporary justice that applies scientific methods and ideas to reflect on the physical evidence related to various criminal cases. Forensic science mostly depends on analytical chemistry to unveil the numerous threads hidden within the trace materials, biological samples, and unknown substances discovered at crime scenes. Among numerous other analytical devices, mass spectrometry (MS) has been the most significant equipment used by forensic scientists to perform the analysis of trace evidence which is distinguished by its high sensitivity, specificity, and versatility. This chapter will take a close look at how MS can be used in different forensic investigations, including toxicology, drug identification, explosive and fire debris, environmental forensics, and questioned document examination. Besides, this chapter will focus on state-of-the-art technologies such as ambient ionization MS, high-resolution MS, and tandem MS, which not only boost the detection and identification of the so-far unknown illegal substances, but also allow for the detection of the same as poisons and trace materials. The chapter will be discussed MS in ink and paper analysis by forensic document examination that will give more understanding of the detection of document forgery. This chapter is designed to give you a clear insight into the development of techniques in the field of criminal investigations of MS, where emphasizing recent technological advances will be our main focus.

**Keywords:** forensic science, forensic chemistry, toxicology, evidence, mass spectrometry

## 1. Introduction

Forensic sciences constitute a multidisciplinary field in which medicine, social sciences, and natural sciences are utilized within the judicial system to elucidate crimes and ensure the prosecution of offenders. The term "forensic sciences," originally "forensic sciences" in English, is etymologically derived from Latin. When the term is analyzed by separating it into its components, the first word, "forensic," originates from the Latin word forensis. This word, in turn, is derived from forum, meaning a public space where, in Ancient Rome, public affairs were discussed, legal cases were heard, and judicial proceedings were conducted. Consequently, forensis carries the meaning of "pertaining to the public." Over time, the term evolved to be used in the sense of "related to court or legal matters" and, in contemporary English, "forensic" has become associated with scientific methods and techniques that can be presented as evidence in court. The second component, "sciences," is derived from the Latin word scientia, which means "knowledge" or "science." "Science" thus refers to the systematic investigation and accumulation of knowledge within a particular field of study [1].

Forensic sciences is a multidisciplinary field that applies scientific methods systematically in crime and crime scene investigations. Analytical instruments used in this field are of great significance for the scientific evaluation and reliable interpretation of physical evidence. Mass spectrometry (MS), one of these instruments, has become one of the primary analytical methods in almost all specialized areas of forensic sciences. Its ability to analyze with high precision, especially at the molecular level, makes it indispensable in both quantitative and qualitative studies [2].

The foundations of mass spectrometry were established in the early twentieth century when J. J. Thomson was reported to separate gaseous ions by mass in 1912. The first commercial mass spectrometers developed in the 1940s were primarily used in physics and chemistry, and later integrated into disciplines such as biochemistry, pharmaceutical research, and environmental analysis [3, 4]. In 1970, Althaus et al. from MIT used computer-assisted GC-MS to detect Darvon (propoxyphene) and its metabolites in the urine of a suspected overdose victim, one of the first examples of the potential of this technique [5]. In the 1990s, the development of LC-MS/MS systems coupled with liquid chromatography enabled the sensitive analysis of highly polar and thermally unstable compounds.

Mass spectrometry is one of the most widely used advanced analysis techniques in forensic science, and this is not only due to its analytical power, but also to the legal credibility it offers. MS instruments are used in many different application areas such as forensic toxicology, explosives and drugs detection, trace evidence

| **Mass spectrometry application in forensic sciences** | |
|---|---|
| Toxicology | • Identifying drugs, poisons, and their metabolites in biological samples (blood, urine, tissues)<br>• Testing for alcohol, drugs in drivers, athletes, workers<br>• Detecting substances used in assaults (e.g., date rape drugs) |
| Drug analysis | • Identifying illegal drugs (heroin, cocaine, methamphetamine) and cutting agents<br>• Detecting synthetic cannabinoids, cathinones, etc. |
| Explosives and fire debris analysis | • Identifying components of explosive<br>• Analyzing accelerants in fire debris (e.g., gasoline residues) |
| Trace evidence analysis | • Chemical characterization of tiny fragments left at crime scenes (paint, fibers, and glass)<br>• Detection of inorganic (lead, barium, antimony) and organic residues from firearms |
| Document examination | • Dating and sourcing inks based on their chemical profiles<br>• Identifying paper type or origin. |
| Environmental analysis | • Tracing toxic chemicals involved in environmental crimes (illegal dumping, poisoning) |
| Forensic anthropology | • Determining diet, migration, or cause of death through isotope ratios |

**Table 1.**
*Application areas of MS in forensic sciences.*

analysis (glass, paint, fiber), document analysis, environmental forensic analysis, and biomarker detection. In addition, in complex matrices, which are often encountered in forensic cases, mass spectrometry enables selective separation of target components and detection of even low concentrations of substances [4]. The application of MS in forensic sciences is summarized in **Table 1**.

In recent years, the development of portable MS systems, ambient ionization techniques (e.g. DESI, DART), and AI-assisted data interpretation software has made MS technologies available even for crime scene analysis and high-speed screening [6, 7]. These developments have enabled MS to be effectively incorporated not only in the laboratory environment but also as a direct part of forensic investigations.

## 2. Mass spectrometry in forensic toxicology and drug analysis

Forensic toxicology analyzes drugs, poisons, and their metabolites in body fluids and tissues, enabling the determination of causes of death, the investigation of substance abuse cases, and the investigation of crime-related chemical exposures. This field has a critical role at the intersection of pharmacology, chemistry, and forensic sciences and plays a critical role in both clinical and legal investigations. Biological samples analyzed in forensic toxicology depend on the way the substance was ingested, the time of ingestion, and the purpose of the investigation. Biological samples that used forensic analysis are given in **Figure 1**.

Blood and urine are the most commonly used matrices for the detection of acute poisonings and recent substance usage. Hair generally uses for retrospective analysis of long-term substance consumption. Hair samples can show substance use up to months or even years ago. Saliva is a non-invasive method for the detection of recent substance use. It is especially preferred in driver inspections. Tissue samples are especially used in postmortem analysis. Organ tissues such as liver and kidney provide information about the distribution of the substance. Alternative matrices include sweat, meconium, breast milk, and vitreous humor. These matrices can provide valuable information, especially when blood and urine samples cannot be obtained [8].

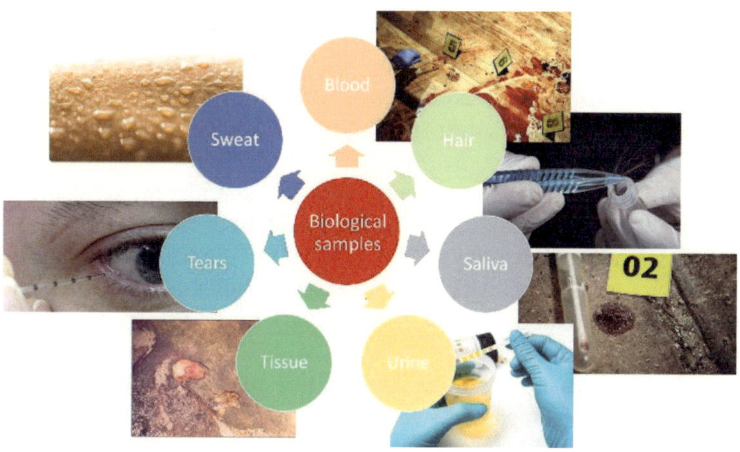


**Figure 1.**
*Biological samples that are used for forensic analysis.*

| Classification of drugs and illicit substances | |
|---|---|
| Stimulants | Amphetamines, methamphetamines, cocaine |
| Depressants | barbiturates, benzodiazepines, GHB, Rohypnol |
| Hallucinogens | LSD, MDMA, PCP, ketamine |
| Synthetic drugs | Bath salts (cathinones), DXM |
| Inhalants | Ether, nitrous oxide, butane, cyclohexyl nitrite, amyl nitrite |
| Narcotics | Opium, heroin, morphine, methadone, oxycodone |

**Table 2.**
*Drug classification in forensic science.*

Generally, drugs examined in forensic toxicology are categorized into the below-mentioned groups and the detection of these drugs is of crucial significance in terms of both legal aspects and public safety. Drug classification is given in the **Table 2**.

These chemicals are detected and quantified using a broad range of mass spectrometry (MS) techniques. These methods are chosen based on the sample matrix, the purpose of the analysis, and the characteristics of the substance to be investigated.

GC-MS has been applied for many decades, particularly for the analysis of classical drugs such as cannabis (THC metabolites), cocaine (and benzoylecgonine), amphetamines, and opiates. However, in recent years, liquid chromatography-tandem mass spectrometry (LC-MS/MS) has been increasingly selected for the analysis of more polar, thermally unstable or high-molecular-weight compounds (e.g., many new psychoactive substances (NPS), benzodiazepines, fentanyl analogs, synthetic cannabinoids). LC-MS/MS generally requires less sample preparation and offers high sensitivity and selectivity [9].

In particular, distinguishing isomeric compounds is a major challenge in forensic toxicology. With the emergence of NPSs such as synthetic cannabinoids or fentanyl analogs, it has become critical to identify a large number of isomers that are structurally similar but have different legal status and potency. Standard GC-MS or LC-MS/MS methods sometimes fail to distinguish isomers. In these cases, chiral chromatography columns or more sophisticated MS techniques (e.g. ion mobility spectrometry-MS or different fragmentation models) may be employed. In cannabis analysis, distinguishing isomers with different legal status, such as delta-9-THC from delta-8-THC, is also a current challenge, and although MS analyses can provide useful information such as characteristic fragment ions, this discrimination is still challenging [10].

LC-MS/MS and GC-MS have been employed in post-mortem toxicological investigations to establish conclusive diagnoses, especially in cases of deaths brought on by polypharmacy (multiple drug usage) or medication combinations that interact with alcohol. In this regard, LC-MS/MS offered comprehensive quantification of both medications and their metabolites in one investigation including fatal cases brought on by a combination of benzodiazepines and opioids [11].

The Michael Jackson case gives an example of this analysis. On June 25, 2009, pop music legend Michael Jackson passed away while receiving medical attention by his personal doctor, Dr. Conrad Murray, and the news was extensively publicized worldwide [12]. According to post-mortem toxicological examination, Jackson's blood and urine contained high levels of benzodiazepines (lorazepam, midazolam) and propofol

(Diprivan). The GC-MS and LC-MS/MS methods were used to analyze drugs [13, 14]. Dr. Murray's negligent medical treatments were established as a consequence of the legal action. He was convicted of homicide in 2011 and given a prison sentence of 4 years.

## 3. Mass spectrometry in explosives and fire debris analysis

Explosive analysis and the investigation of arson incidents are one of the most challenging and important aspects of the forensic sciences. In such incidents, obtaining reliable scientific evidence of the origin of the fire or explosion is critical to the integrity of the forensic process and public safety [15]. Mass spectrometry (MS) is one of the most powerful analytical techniques used in this field. In particular, GC-MS coupled with gas chromatography and LC-MS/MS coupled with liquid chromatography enable highly sensitive detection of both explosive residues and ignitable liquids left behind after a fire [16]. With these methods, even microscopic amounts of evidence from the scene can be accurately identified and the cause of the incident can be elucidated.

In the 1960s, Joseph Nicol of the Chicago Police Crime Laboratory proposed the use of GC for the analysis of volatile organic compounds in fire residues in arson cases [17]. However, when the data obtained by GC were insufficient in complex sample matrices, in the mid-1970s Zoro and Hadley found that GC-MS was more selective than GC-FID. This system enabled both the separation and identification of volatile compounds, increasing the precision of the identified compounds. Subsequently, the application of GC-MS systems has increased [18]. In the 1980s, the National Institute of Standards and Technology (NIST) and the American Society for Testing and Materials (ASTM) pioneered the standardization of laboratory practices for arson analysis and GC-MS became the main reference method [19]. In particular, the ASTM E1618 standard established the systematic use of GC-MS for the identification of ignition liquid residues. In the same years, the use of GC-MS in explosives analysis also increased, making it possible to reliably detect the residues of high explosives such as TNT, RDX, and PETN [20]. Since the 1990s, the development of tandem mass spectrometry (MS/MS) and high-resolution MS systems (HRMS) has made it possible to detect compounds at much lower concentrations. Today, time-of-flight mass spectrometry (TOF-MS) coupled with two-dimensional gas chromatography (GC × GC) offers detailed analysis of even complex fire residues [21]. **Figure 2** shows the timeline of evolution.

LC-MS/MS systems provide critical benefits in the analysis of explosive compounds with low ionization efficiency or non-volatile explosive compounds. They are especially preferred in forensic analysis in terms of reducing ion suppression in liquid samples and achieving very low detection limits. There are many studies on LC-MS/MS applications in explosive and fire debris analysis in the literature.

In 2021, the study investigates the applicability of the triple quadrupole liquid chromatography-mass spectrometry (LC-MS/MS) technique to identify triglycerides in fire and liquid residues. The researchers tested various fat samples in controlled fire scenarios to determine whether animal and vegetable fats leave detectable traces in post-combustion residues. The LC-MS/MS system selectively detected specific molecular ions and breakdown products of triglycerides, allowing the identification of these components even in complex matrices. The results show that the presence of triglycerides can help to determine the type of materials or accelerants used in fire initiation, thus emphasizing that the method can be used as both a descriptive and a classification instrument in forensic fire analysis [22].

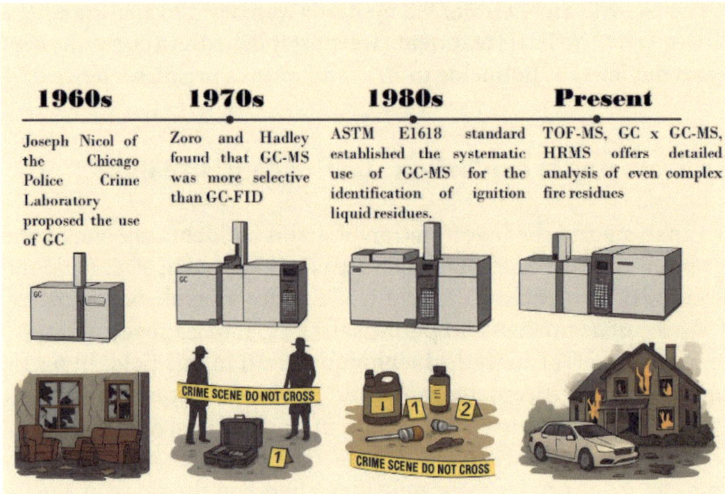


**Figure 2.**
*Key milestones of mass spectrometry in explosive and fire debris analysis.*

In 2023, Spagnesi et al. combined continuous flow analysis (CFA) with fast liquid chromatography and tandem mass spectrometry (FLC-MS/MS) techniques for the detection of biomass combustion markers such as levoglucosan in ice cores from the Swiss Alps. The developed method detects levoglucosan with a detection limit as low as 66 ng/L, enabling temporal and regional monitoring of past fire events. This approach is considered as an important step in determining fire history with high precision [23].

IRMS provides unique information for both source identification and production process monitoring by analyzing the isotopic signatures of chemical compounds. Comparing the isotope ratios of explosives enables the determination of whether suspected substances come from the same source.

The 2009 study by Sarah J. Benson et al. is a pioneering study evaluating the potential of isotope ratio mass spectrometry (IRMS) in forensic explosives analysis. The research focused on the analysis of light stable isotopes (carbon, hydrogen, oxygen, and nitrogen) in order to overcome the limited discrimination of traditional forensic techniques. In particular, it examined whether samples of explosives such as tri-acetone tri-peroxide (TATP) and pentaerythritol tetranitrate (PETN) from different sources could be distinguished. In the study, carbon isotope values of 14 TATP samples were measured and these data allowed the samples to be classified into three distinct groups. In addition, the carbon and hydrogen isotope compositions were used separately and in combination with oxygen isotope values in the analysis of four additional TATP samples to clearly differentiate the samples. The 3D plot of these three isotopes showed that the TATP samples were clearly distinguishable. In the analysis of PETN samples, carbon and nitrogen isotope values of 15 samples were measured and it was determined that samples from different sources could be easily separated. These findings indicate that IRMS is a successful application in forensic explosives analysis to distinguish samples from different sources. The research represents a preliminary evaluation of the IRMS technique for the measurement of stable isotope values in TATP and PETN samples and supports funding for a full evaluation of this application [24].

The study by Gentile et al. published in 2011 provides a methodological approach to the use of isotope ratio mass spectrometry (IRMS) in forensic science. The paper discusses how IRMS can be an effective tool for source identification of materials such as explosives, drugs, and biological samples using carbon-13 and nitrogen-15 isotopes. Experimentally, samples were properly collected and prepared, IRMS instruments were calibrated, isotopic ratios were measured, and data were processed using normalization techniques. The results obtained showed that IRMS offers a reliable method to determine the origin of samples by isotopic analysis [25].

On December 21, 1988, Pan Am Airlines Flight 103 from London to New York exploded over Lockerbie, Scotland. In the incident, 259 people on board and 11 people on the ground lost their lives. The incident went down in history as one of the worst terrorist attacks on British soil [26]. Detailed analysis of the debris collected after the explosion revealed explosive residues on a small piece of plastic stuck in a piece of luggage. The explosive was identified as Semtex plastic type. Gas chromatography-mass spectrometry (GC-MS) and ion mobility spectrometry were used to detail the chemical components of the explosive. The presence of RDX and PETN in the explosive and the ratios of the trace compounds provided critical information about the production method of the bomb used [27–29]. Further analysis revealed that the explosive was supplied by a group linked to the Libyan government. As a result of these scientific evidence, the Libyan agent Abdelbaset al-Megrahi was convicted and sentenced to life imprisonment after a trial in the Netherlands in 2001 [30].

## 4. Mass spectrometry in trace evidence analysis

Trace evidences are physical materials found at a crime scene, usually in micro-quantities. This evidence can play a critical role in establishing a link between a person or object and a crime scene or victim. Through its high sensitivity and ability to providc information on chemical structure, mass spectrometry (MS) is a powerful tool for analyzing a broad range of trace evidence, from paint residue to fibers, and glass fragments to gunshot residue. **Figure 3** shows that the application of mass spectrometry in trace evidence analysis.

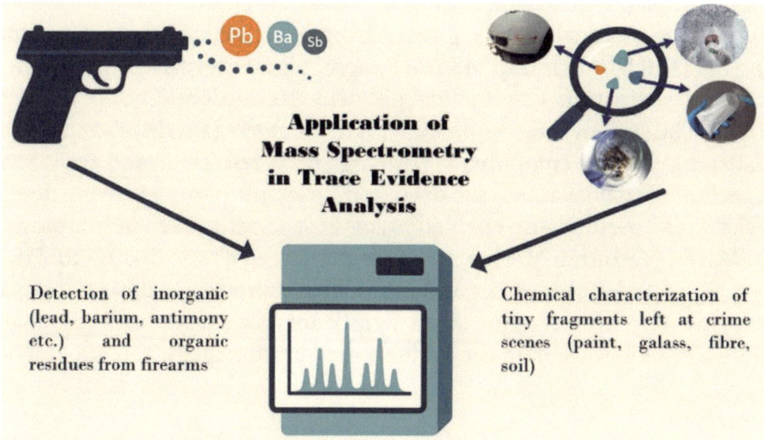


**Figure 3.**
*Application of mass spectrometry in trace evidence analysis.*

### 4.1 Glass analysis

Fragments of glass are among the most frequently encountered trace evidence at crime scenes. Broken glass fragments can be transferred from a crime scene to a suspect's clothing or vehicle, and are therefore critical for the investigation of a crime. The materials and additives used in the glass production process give each piece of glass a unique elemental profile. Mass spectrometry analysis of these elements (e.g., strontium, zirconium, barium, rare earth elements) plays a critical role in distinguishing glass sources (e.g., window glass, car headlights, bottles), determining whether the particles have a common origin, and connecting them to the crime scene [31]. LA-ICP-MS and LA-ICP-TOF-MS techniques used in glass analysis simultaneously identify a large number of elements in a short time and determine the similarities and differences between different glass samples. This provides an advantage in forensic analysis. In 2023, Becker and Günther used the LA-ICP-TOF-MS technique to reduce the amount of sample required for forensic glass analysis. This study demonstrated that even very small glass fragments can be reliably analyzed with the single-shot ablation method. Furthermore, the integration of multivariate statistical analysis further strengthened the discrimination between different glass samples [32].

### 4.2 Paint analysis

Paint residues are among the most common trace evidence, especially in incidents such as traffic accidents, arson, and vandalism. Analyzing paint samples is critical for identifying the sources of materials at the scene and understanding how the crime was committed. These analyses play an important role in linking the suspect to the incident and ensuring that the evidence can be presentable in court. Although many different methods are used in paint analysis, mass spectrometry is the most preferred analytical technique in forensic paint analysis because it determines the chemical composition of paint samples with high sensitivity and specificity. In particular, it is used to identify the organic and inorganic components of dyes, to distinguish between different dye samples and to identify their possible sources. This technique enables fast and efficient analysis with minimal sample size [33].

Chen and Wu investigated the potential of using Q-Orbitrap tandem mass spectrometry (MS/MS) coupled with a Direct Real-Time Analysis (DART) ionization source for rapid, direct, and *in situ* analysis of pigments from paint evidence in forensic applications. Using the DART-Q-Orbitrap MS/MS system, pigments from paint samples were analyzed without extraction. Commercial pigment standards and real paint samples were analyzed. The system was combined with tandem MS (MS/MS) capabilities to elucidate the structures of the components with high mass resolution and accuracy. Many pigments, including thalocyanins, azo dyes, and inorganic pigments, were detected with high specificity. The method also enabled pigment separation even in complex dye mixtures. The DART-Q-Orbitrap MS/MS system has emerged as an effective tool in forensic science, enabling rapid, non-contact and minimal preparation analysis of pigments from paint evidence. This technique offers significant advantages in on-site analysis, such as crime scene analysis, or in the evaluation of limited quantities of evidence [34].

### 4.3 Fiber analysis

Fibers are among the most frequently collected trace evidence at crime scenes. Particularly textile fibers play an important role in establishing a link between the suspect and the victim or crime scene. Characteristics of fibers such as chemical composition, pigment type, and production process are used for source identification and comparative analysis. Mass spectrometry is often the method of choice to analyze the polymeric structure of the fiber (e.g. nylon, polyester, acrylic) and sometimes the additives it contains. In particular, it is used for the identification of colorants and additives in fibers, to distinguish between different fiber samples and identify their possible sources. This technique allows for fast and efficient analysis with minimal sample size. Pyrolysis-GC-MS (Py-GC-MS), LC-MS, and MALDI-MS are generally used in fiber analysis [35].

In 2019, Liang et al. investigated the feasibility of Direct Real-Time Analysis-Mass Spectrometry combined with Thermal Desorption/Pyrolysis (TD/Py-DART-MS) for the rapid and efficient analysis of fiber samples encountered in forensic investigations. Different types of synthetic and natural fibers (e.g. polyester, nylon, acrylic, cotton) were analyzed. The fiber samples were directly heated (thermal desorption and pyrolysis) to release volatile components and the vapor phase was analyzed by DART-MS. The mass spectra obtained revealed the chemical signatures of the fiber species. Both polymer parent structures and additives (e.g. dyes, plasticizers) were determined from fiber samples by TD/Py-DART-MS. It was also found that a very small amount of fiber material (e.g. 1–2 mm yarn fragments) is sufficient for analysis, and the analysis time is very relatively fast and does not require sample preparation [36].

### 4.4 Soil analysis

Soil is common trace evidence that can be easily transferred from a crime scene to a suspect's shoes, clothing, or vehicle. The analysis of soil samples plays a critical role in determining whether the suspect was present at the crime scene. Soils in different geographical locations differ in both inorganic (mineral) and organic (humus, plant residues, contaminants) compositions, and these characteristics are used for source identification in forensic analysis. Although there are many spectrometric methods (e.g., laser-induced breakdown spectroscopy (LIBS)) used to determine the elemental composition of soil samples, mass spectrometry is often preferred in forensic soil analysis due to its high sensitivity and specificity. In this regard, GC-MS and LC-MS are employed for quantifying organic molecules in soil (such as pesticides, hydrocarbons, and plant biomarkers), whereas ICP-MS is used to identify trace element profiles in soil [37].

In a comparative forensic analysis of soil samples, the pyrolysis-gas chromatography/mass spectrometry (Py-GC/MS) technique was used to characterize the organic matter composition at the molecular level. Twelve soil samples from different regions in Brazil were studied and analyzed at the molecular level by Py-GC/MS to separate the organic compounds in the soil samples. The soil organic matter varied in terms of characteristic compounds such as lignin, lipids, proteins, polysaccharides, and alkylbenzenes. The study revealed that the pyrolysis-gas chromatography/mass spectrometry (Py-GC/MS) technique can be used as an effective tool for matching soils to crime scenes, especially when they are carried on the suspect's shoes, vehicle, or clothing [38].

Sixteen years ago, a person went missing in the United Kingdom and was never found despite extensive searches. Years later, the area where the missing person was buried was identified after a search in line with the intelligence gathered. In the study by Donnelly, Cassella et al. [39], mass spectrometry played a critical role in the identification of organic compounds in forensic soil analysis carried out in the area where the body was found within the scope of this murder case. In soil samples taken from the vicinity of the grave, putrescine, and similar polyamines, biomarkers specific to the decomposition process (directly related to corpse decomposition and concentrated only in burial sites), stanols (organic substances associated with human excrement or bodily decomposition processes; e.g., coprostanol, cholestanol), and other organic decomposition products were successfully detected by mass spectrometry, confirming that the burial site was exposed to biological activity. These analyses revealed not only the presence and location of the body, but also biochemical traces of environmental changes following the burial process. Mineralogical analysis, integrated with other techniques, established a multidimensional link between the crime scene and the body. The study demonstrated the power of mass spectrometry as a robust and complementary technique not only for biomarker detection, but also for forensic time estimation and crime scene reconstruction [39].

### 4.5 Gunshot residue (GSR) analysis

When a firearm is discharged, small particles from the weapon and ammunition are scattered in the environment. These GSR can be found on the shooter's hands, clothing or on nearby surfaces. Traditionally, GSR analysis is performed by scanning electron microscopy-energy dispersive X-ray spectroscopy (SEM-EDX) and is based on morphological and elemental analysis of characteristic spherical particles containing lead (Pb), antimony (Sb), and barium (Ba). However, MS, in particularly ICP-MS, offers complementary or alternative approaches for GSR analysis. ICP-MS can measure a large number of elements (e.g., bullet or shell components such as tin, copper, zinc, aluminum) with higher sensitivity than SEM-EDX. Laser ablation-ICP-MS (LA-ICP-MS) can be used to map the elemental composition of individual particles or the GSR distribution on surfaces. Furthermore, isotope ratio analyses by ICP-MS (e.g., lead isotopes) offer the potential to discriminate between different ammunition sources, increasing the discrimination of GSR evidence [40].

The study published by Ledergerber et al. [41] evaluated the applicability of direct analysis real-time mass spectrometry (DART-MS) on organic gunshot residues (OGSR) on a series of different surfaces. In the study, 330 samples from different surfaces were analyzed and DART-MS was shown to be a fast and effective method for OGSR detection. In particular, the detection limits of DART-MS ranged from 0.075 to 12 ng, with intra- and inter-day repeatability below 0.0012%. This demonstrates that the method is both sensitive and reliable. Furthermore, the mass spectra obtained at different in-source collision-induced dissociation (is-CID) voltages enabled the separation of isomers. However, the percentage of true positives in samples taken from the shooter's hands ranged from 19 to 73%, depending on the type of ammunition and the classification criteria. This suggests that DART-MS is a promising method for OGSR detection [41].

Another study conducted in 2024 investigated the feasibility of GSR detection in evidence materials containing bloodstains by inductively coupled plasma-mass spectrometry (ICP-MS). In the study, GSR analyses were performed on both dry and fresh bloodstained clothing and cotton samples, and it was demonstrated that classical inorganic GSR elements such as lead (Pb), barium (Ba), and antimony (Sb) can be reliably detected even at extremely low levels (ng levels) by ICP-MS. Despite the matrix effect of blood, the high sensitivity and multi-element analysis capability of the ICP-MS enabled quantitative measurement of GSR components and signal suppression due to biological matrix was minimized by optimizing sample preparation and dilution steps. The study also evaluated the effects of contamination risk and sample acquisition timing on GSR detection, highlighting the importance of time-sensitive analysis protocols to maintain the forensic validity of such traces. In conclusion, it was demonstrated that ICP-MS can be used as a powerful analytical tool for GSR analysis not only in clean laboratory samples but also in complex biological samples, and that this approach can be integrated into GSR examinations in bloody clothing or body samples obtained from armed incidents. This study demonstrates the forensic potential of advanced analytical methods to overcome biological barriers in GSR analysis [42].

## 5. Mass spectrometry in questioned document analysis

The analysis of questioned documents (forged coins, wills, contracts, etc.) is an important field of forensic sciences. Mass spectrometry analysis of the chemical structure of the paper and ink, which are the constituent elements of the document, is an important method used to determine the time of creation of the document. GC-MS, TD-GC/MS methods are frequently used to analyze the solvents in the ink structure to determine the creation time of documents between 0 and 12 months [43]. Laser ablation-ICP-MS (LA-ICP-MS) can be used to analyze inorganic element profiles in inks or paper [44], while LC-MS and MALDI-MS are also used for more detailed structural analysis of ink components [45]. More advanced techniques such as Direct Analysis in Real Time (DART-MS) and Desorption Electrospray Ionization (DESI-MS) can rapidly analyze ink components (pigments, dyes, solvents, binders) by sampling directly from the surface of the document. This allows to discriminate between inks from different pens or printers, to determine if a document has been added to, or to comment on the age of the ink (by comparing it to artificial aging studies) [46]. **Figure 4** shows application of mass spectrometry in questioned document analysis.

Costa, Karina F.F., and colleagues present an innovative approach that combines direct injection mass spectrometry (DI-MS) and correlation analysis to estimate the age of ballpoint pen ink writing in documents. In this study of 505 ink samples

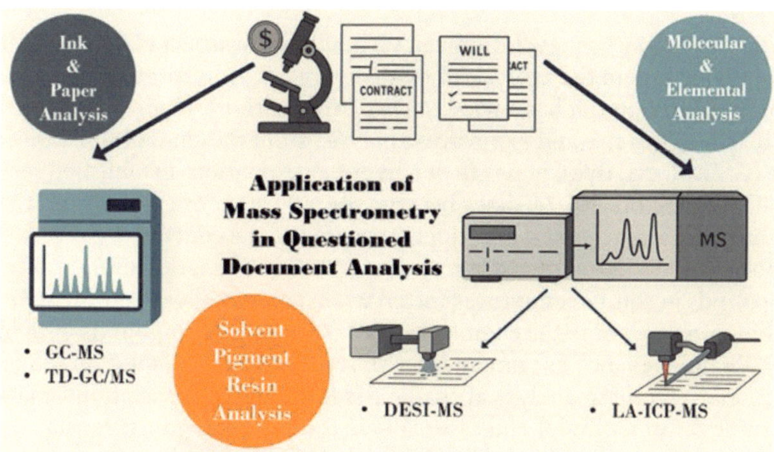


**Figure 4.**
*Applications of mass spectrometer in questioned document analysis.*

collected between 1962 and 2014 from the archives of the National Institute of Criminalistics of the Brazilian Federal Police, DI-MS was used to understand how ink components degrade over time. The mass spectra were evaluated based on normalized peak areas and correlation analysis was performed using the Unsupervised Linkage Threshold (ULT) method using the cosine square function as a similarity measure. This analysis allowed reliable age estimates and correlation thresholds to be determined for different ink formulations. In particular, degradation pathways such as the transformation of the commonly used Crystal Violet (CV) dye into pararosaniline through demethylation over time and the formation of by-products such as N,N′-Dimethyl-4-aminophenol (NNAPH) were investigated. These degradation products were observed to exhibit similar time profiles in certain ink brands. The developed method is of great advantage in forensic applications as it can estimate ink age without being dependent on factors that are difficult to know *a priori*, such as storage conditions or initial ink composition. This study opens a new avenue for estimating the date of writing of documents using the chemical degradation profiles of ink components and shows that consistent age estimates can be made even across different ink brands [47].

Research in 2022 examined the applicability of desorption electrospray ionization mass spectrometry (DESI-MS) imaging to determine the age of ink writing on documents. In the study, blue ballpoint pen and black gel ink samples were artificially aged by heat and light exposure and then analyzed by DESI-MS. During the analysis, a methanol-water-formic acid mixture (90,10,0.1%, v/v/v) was used as desorption solution. In the chemical images obtained, significant differences in the distribution of age-related compounds such as basic dyes and polyethylene glycol derivatives were observed. It was also found that the amount of ethoxylated fatty amine surfactants present in the copier paper decreased after aging. This method made it possible to visualize the compositional differences between naturally aged and fresh ink samples and demonstrated that analysis can be performed without damaging the samples. In conclusion, the study shows that the DESI-MS imaging technique is a promising method for determining the relative age of ink writing on documents [48].

## 6. Conclusions

Mass spectrometry has revolutionized virtually every aspect of forensic science, providing advantages in the analysis of evidence and the interpretation of results. As discussed in this chapter, it has a wide range of applications, from drug detection in forensic toxicology to fire and explosives analysis, from the characterization of trace evidence (GSR, fibers, dyes, glass) to document examination. Established techniques such as GC-MS and LC-MS/MS have become the gold standard for routine analysis in many forensic laboratories and are widely accepted in the courts.

The power of MS in forensic science relies on the combination of its ability to separate compounds in complex mixtures (often when combined with chromatography) and its ability to determine the chemical structure of these compounds with high certainty. Techniques such as tandem MS (MS/MS) and high-resolution MS (HRMS) are critical in overcoming analytical challenges (e.g., low concentrations, matrix effects, isomeric compounds) and especially in the struggle against constantly emerging threats such as new psychoactive substances (NPS). ICP-MS has revolutionized trace element analysis, while IRMS offers valuable insights in source identification and counterfeit detection.

However, despite the tremendous potential offered by MS technology, some limitations and challenges are still remained. Some of these challenges are the high cost, the need for specialized personnel to operate and maintain sophisticated instruments, time-consuming method development and validation, complex data interpretation, and lack of standardization.

In the future, the role of mass spectrometry in forensic science is expected to grow even further. Technological advances are leading to more sensitive, faster and more user-friendly instruments. Techniques such as DART, DESI, and LAESI have the potential to shorten screening times at the crime scene or in the laboratory by enabling rapid analysis directly from surfaces or objects with minimal or no sample preparation. It can be said that their use will increase in areas such as document examination, explosive detection, and drug screening. With advances in technology, the development of more compact, robust and portable MS devices that can be used directly in the field will enable rapid and preventive analysis at the scene. Machine learning and artificial intelligence tools are expected to play an important role in automating data analysis and pattern recognition, as advanced chemometric analyses are needed to process, interpret, and draw meaningful conclusions from the large amounts of MS data generated.

In conclusion, mass spectrometry is a fundamental technology that has shaped the past and will continue to shape the future of forensic science. As technological advances continue, the contribution of MS to forensic investigations will only increase, but responsible use of this powerful tool, rigorous validation of methods, careful interpretation of results, and compliance with legal standards will always be a priority.

## Conflict of interest

The authors declare no conflict of interest.

## Author details

Dilek Salkım İşlek* and Eda Kiriş
Istanbul University-Cerrahpasa Institute of Forensic Sciences and Legal Medicine, Istanbul, Turkey

*Address all correspondence to: dsislek@iuc.edu.tr

## References

[1] Houck MM, Siegel JA. Fundamentals of Forensic Science. Oxford, UK: Academic Press; 2009

[2] Gross JH. Mass Spectrometry: A Textbook. 2nd ed. Cham, Switzerland: Springer; 2011

[3] Sparkman OD, Penton Z, Kitson FG. Gas Chromatography and Mass Spectrometry: A Practical Guide. 2nd ed. Oxford, UK: Academic Press; 2011

[4] Cooks RG, Ouyang Z, Takats Z, Wiseman JM. Ambient mass spectrometry. Science. 2006;**311**(5767):1566-1570

[5] Althaus JR, Biemann K, Biller J, Donaghue PF, Evans DA, Förster HJ, et al. Identification of the drug Darvon and its metabolites in the urine of a comatose patient using a gas chromatograph-mass spectrometer-computer system. Experientia. 1970;**26**(7):714-717

[6] Jaitly N, Monroe ME, Petyuk VA, Clauss TR, Adkins JN. Robust algorithm for alignment of LC-MS analyses in an accurate mass and time tag pipeline. Analytical Chemistry. 2006;**78**(21):7397-7409

[7] Nemes P, Vertes A. Laser ablation electrospray ionization for in vivo and imaging mass spectrometry. Analytical Chemistry. 2007;**79**(21):8098-8106

[8] Dinis-Oliveira RJ, Carvalho F, Duarte JA, Remiao F, Marques A, Santos A, et al. Collection of biological samples in forensic toxicology. Toxicology Mechanisms and Methods. 2010;**20**(7):363-414

[9] Brown HM et al. The current role of mass spectrometry in forensics and future prospects. Analytical Methods. 2020;**12**(32):3974-3997

[10] Reber JD, Karschner EL, Seither JZ, Knittel JL, Dozier KV, Walterscheid JP. An enhanced LC-MS-MS technique for distinguishing Δ8- and Δ9-tetrahydrocannabinol isomers in blood and urine specimens. Journal of Analytical Toxicology. 2022;**46**(4):343-349. DOI: 10.1093/jat/bkac007

[11] Hathaway RA et al. Fatal polydrug use involving opioids and benzodiazepines: A postmortem toxicological investigation. Journal of Analytical Toxicology. 2021;**45**(3):237-245. DOI: 10.1093/jat/bkaa078

[12] Wehrwein P. Propofol: The drug that killed Michael Jackson. Harvard Health Publishing. 2011. Available from: https://www.health.harvard.edu/blog/propofol-the-drug-that-killed-michael-jackson-201111073772 [Accessed: May 03, 2025]

[13] Maas A, Maier C, Iwersen-Bergmann S, Madea B, Hess C. Simultaneous extraction of propofol and propofol glucuronide from hair followed by validated LC-MS/MS analyses. Journal of Pharmaceutical and Biomedical Analysis. 2017;**146**:236-243

[14] Vaiano F, Serpelloni G, Focardi M, Fioravanti A, Mari F, Bertol E. LC-MS/MS and GC-MS methods in propofol detection: Evaluation of the two analytical procedures. Forensic Science International. 2015;**256**:1-6

[15] Stauffer E, Dolan JA, Newman R. Fire Debris Analysis. London, UK: Academic Press; 2007

[16] Sharma B, Gadi R. Analytical tools and methods for explosive analysis in forensics: A critical review. Critical Reviews in Analytical Chemistry. 2025;**55**(2):251-277

[17] Nicol JD. Police science technical abstracts and notes. Safety. 1949;**35**:6-7

[18] Zoro JA, Hadley K. Organic mass spectrometry in forensic science. The Journal of Forensic Sciences. 1976;**16**(2):103-114

[19] Smith RM. Arson analysis by mass chromatography. Analytical Chemistry. 1982;**54**(13):1399A-1409A

[20] American Society for Testing and Materials. ASTM E1618-14: Standard test method for ignitable liquid residues in extracts from fire debris samples by gas chromatography-mass spectrometry; 2014

[21] Nizio KD, Cochran JW, Forbes SL. Achieving a near-theoretical maximum in peak capacity gain for the forensic analysis of ignitable liquids using GC×GC-TOFMS. Separations. 2016;**3**(3):26. DOI: 10.3390/separations3030026

[22] Bryant CM et al. Identification of triglycerides in liquid and fire debris samples by triple quadrupole liquid chromatography-mass spectrometry. Journal of Forensic Sciences. 2021;**66**(2):534-546

[23] Spagnesi A et al. An upgraded CFA–FLC–MS/MS system for the semi-continuous detection of levoglucosan in ice cores. Talanta. 2023;**265**:124799

[24] Benson SJ et al. Forensic analysis of explosives using isotope ratio mass spectrometry (IRMS)—Preliminary study on TATP and PETN. Science and Justice. 2009;**49**(2):81-86

[25] Gentile N et al. On the use of IRMS in forensic science: Proposals for a methodological approach. Forensic Science International. 2011;**212**(1-3):260-271

[26] Federal Bureau of Investigation (FBI). 30 Years Later, Still Actively Seeking Justice. 2018. Availablefrom: https://www.fbi.gov/news/stories/remembering-pan-am-flight-103-30-years-later-121418 [Accessed: May 07, 2025]

[27] Global Forensic and Justice Center-GFJC. A Simplified Guide to Forensic Science, Explosives. 2013. Available from: https://www.forensicsciencesimplified.org/explosives/how.html [Accessed: May 07, 2025]

[28] Moore S, Schantz M, MacCrehan W. Characterization of three types of semtex (H, 1A, and 10). Propellants, Explosives, Pyrotechnics. 2010;**35**(6):540-549

[29] Vermeij E. Survivability of IED components, suitcases, their contents and luggage containers in suitcase bombs. Forensic Science International: Reports. 2024;**9**:100366

[30] The Editors of Encyclopaedia Britannica. Pan Am Flight 103. 2008. Available from: https://www.britannica.com/event/Pan-Am-flight-103 [Accessed: May 07, 2025]

[31] Almirall J, Trejos T, Lambert K. Interpol review of glass and paint evidence 2016-2019. Forensic Science International: Synergy. 2020;**2**:404-415

[32] Becker P, Günther D. Reducing sample amount for forensic glass analysis using LA-ICP-TOFMS and multivariate statistics. Journal of Analytical Atomic Spectrometry. 2023;**38**(8):1704-1712

[33] Lee Y et al. Application of time-of-flight secondary ion mass spectrometry to automobile paint analysis. Analytical Sciences. 2001;**17**(6):757-761

[34] Chen T-H, Shu-Pao W. Forensic applications of direct analysis in real time (DART) coupled to Q-orbitrap

tandem mass spectrometry for the in situ analysis of pigments from paint evidence. Forensic Science International. 2017;**277**:179-187

[35] Lepot L et al. Interpol review of fibres and textiles 2019-2022. Forensic Science International: Synergy. 2023;**6**:100307

[36] Liang J et al. Forensic fiber analysis by thermal desorption/pyrolysis-direct analysis in real time-mass spectrometry. Analytical Chemistry. 2019;**92**(2):1925-1933

[37] Fitzpatrick RW. "Soil: Forensic Analysis." Wiley Encyclopedia of Forensic Science. Chichester: John Wiley and Sons; 2009. pp. 2377-2388

[38] Mazzetto JML et al. Potential of soil organic matter molecular chemistry determined by pyrolysis-gas chromatography/mass spectrometry for forensic investigations. Science and Justice. 2019;**59**(6):635-642

[39] Donnelly LJ, et al. Analysis of Soil Following a Police-Led Open Area Search and the Recovery of a Cold-Case Homicide Grave. Bath, UK: The Geological Society; 2021

[40] Brożek-Mucha Z. Trends in analysis of gunshot residue for forensic purposes. Analytical and Bioanalytical Chemistry. 2017;**409**:5803-5811

[41] Ledergerber TD et al. A feasibility study of direct analysis in real time-mass spectrometry for screening organic gunshot residues from various substrates. Analytical Methods. 2023;**15**(36):4744-4757

[42] Li X et al. A study on the measurement of GSR with bloodstains by ICP-MS. Forensic Sciences Research. 2024;**9**(4):owad033

[43] Gupta V, Singh G, Saini K. Determination of the age of gel pen inks on handwritten documents using gas chromatography-mass spectrometry and chemometrics. Journal of Forensic Document Examination. 2024;**32**:17-24

[44] Trejos T, Flores A, Almirall JR. Micro-spectrochemical analysis of document paper and gel inks by laser ablation inductively coupled plasma mass spectrometry and laser induced breakdown spectroscopy. Spectrochimica Acta Part B: Atomic Spectroscopy. 2010;**65**(11):884-895

[45] Favaro G et al. Application of LC-MS and LC-MS-MS to the analysis of photo-decomposed crystal violet in the investigation of cultural heritage materials aging. Journal of Mass Spectrometry. 2012;**47**(12):1660-1670

[46] Drury N, Ramotowski R, Moini M. A comparison between DART-MS and DSA-MS in the forensic analysis of writing inks. Forensic Science International. 2018;**289**:27-32

[47] Costa KFF et al. Document ink dye age estimation by direct injection-mass spectrometry and correlation analysis. Microchemical Journal. 2019;**147**:1123-1132

[48] Sun Q et al. Comparative analysis of aged documents by desorption electrospray ionization–mass spectrometry (desi-ms) imaging. Journal of Forensic Sciences. 2022;**67**(5):2062-2072

Chapter 5

# Mass Spectrometry for In-Depth Study and Discovery of Marine Bioactive Metabolites

*Elando Fréda Zamanileha, Marnella Quetty Lema, Antonio Randriampahasoavana, Stanislas Fulgence Lahadson, Picardino Frienduc Vaonalamihanta, Andriamanarivosoa Rija Razafintsalama, Gilles Bedoux and Pierre Hervé Ravelonandro*

## Abstract

This chapter examines the use of mass spectrometry as an effective tool for the detailed study and discovery of bioactive metabolites of marine origin. The seas and oceans are full of organisms that generate a significant variety of natural compounds with pharmacological, cosmetic, and nutraceutical effects. Thanks to its high sensitivity and analytical resolution, mass spectrometry is able to identify, characterize, and quantify these molecules, even when they are found in complex matrices. This chapter discusses the main categories of marine metabolites, methods frequently used in mass spectrometry, and approaches for discovering new bioactive substances. It also highlights contemporary issues and future directions concerning the sustainable use of these natural resources in research and industry.

**Keywords:** bioprospecting, chromatography, marine bioactives, mass spectrometry, metabolites, natural compounds

## 1. Introduction

The oceans, which cover more than 70% of the planet's surface, represent one of the largest reservoirs of biodiversity, with 34 of the 36 major branches of life [1, 2]. Yet, only around 10% of the 150,000 natural metabolites described to date are marine in origin, revealing an immense untapped potential [1, 3]. Life in extreme marine environments, marked by fluctuating salinity, pressure, temperature and light, has driven the production of unique secondary metabolites rarely found in terrestrial species [4]. These compounds play vital ecological roles (defense, communication, and competition) [5] and display diverse bioactivities, including anticancer, antimicrobial, antiviral, anti-inflammatory, and antioxidant properties [6, 7].

Marine bioprospecting targets organisms recognized for their exceptional chemical diversity. Macroalgae (brown, red, and green) produce metabolites such as phlorotannins, carotenoids, and sulfated polysaccharides with antiviral and immunomodulatory properties [8, 9]. Sponges (*Porifera*) have yielded key pharmacological compounds: discodermolide and halichondrin B inspired synthetic anticancer agents, while nucleoside analogs led to cytarabine, a chemotherapy drug [10]. Tunicates produce potent alkaloids, notably trabectedin from *Ecteinascidia*, now clinically approved [11]. In addition, marine microorganisms (cyanobacteria and actinomycetes) cultivated under OSMAC strategies reveal novel antibiotics and anticancer agents. Together, these groups highlight the vast reservoir of bioactive marine metabolites, supporting global bioprospecting efforts [12].

The study of marine metabolites faces analytical challenges, as they are often present at very low concentrations within salt- and pigment-rich matrices [13]. Mass spectrometry (MS) has become a reference technique due to its high sensitivity for detecting trace compounds. When combined with chromatography (LC-MS and GC-MS), it enhances separation and structural analysis. High-resolution instruments and integrated metabolomic strategies with NMR and bioinformatics further improve identification while limiting rediscovery [14].

Recent analytical progress underpins the sustainable exploitation of marine bioactive metabolites. Beyond their pharmaceutical importance, these compounds are applied in cosmetics and nutrition, notably as antioxidants, omega-3 fatty acids, and carotenoids [5, 10]. To protect ecosystems, alternatives such as cultivating producer organisms, biotechnological synthesis, and valorizing fishing by-products are encouraged [15]. Mass spectrometry lies at the core of this strategy, bridging analytical innovation and sustainable development and enabling rational exploration of marine resources [16]. This chapter reviews advances in LC-MS, GC-MS, and metabolomic bioinformatics, illustrating their role in deciphering marine chemistry and supporting industrial, eco-responsible applications.

## 2. Diversity of marine bioactive metabolites and bioprospecting challenges

### 2.1 Major categories of marine bioactive metabolites

Marine secondary metabolites are an exceptional reservoir of chemical and pharmacological innovation. Polyketides play a central role, synthesized by polyketide synthases, showing great structural and functional diversity [17]. One example is bryostatin, isolated from *Bugula neritina*, which modulates protein kinase C and has undergone clinical evaluation for anticancer applications [18]. Cyclic and nonribosomal peptides (NRPs) also stand out: didemnins from the tunicate *Trididemnum solidum* exhibit strong antitumor properties and led to the development of plitidepsin [19]. Halogenated alkaloids, such as fascaplysine from the sponge *Fascaplysinopsis bergquist*, act through CDK4 (cyclin-dependent kinase 4) inhibition, while others display antimicrobial and neuroactive effects [20]. In addition, marine organisms supply specialized lipids like PUFAs (EPA, DHA) crucial for human health [15] and fucoidans from brown algae with antithrombotic, antiviral, and immunomodulatory activities, though their structural complexity complicates analysis (**Table 1**) [29, 30].

| Secondary metabolite | Metabolic pathways | Ecological implications | Ref. |
|---|---|---|---|
| Tetrabromopyrrole (*Pseudoalteromonas*, coralline algae) | Pyrrole halogenation (bromoperoxidases) | Signal for coral larval metamorphosis; toxic at high levels | [21] |
| Halimide (*Aspergillus-Halimeda*) | Diketopiperazine/ polyketide-NRP | Precursor of Plinabulin; fungal defense | [22] |
| Pestalone (*Pestalotia*-brown algae) | PKS activation in co-culture with bacteria | Strong antibacterial activity; algae-fungus protection | [23] |
| Marinomycine A (Marinispora) | Modular PKS macrolide | Cytotoxic to melanoma cells; microbial defense | [24] |
| Salinosporamide A (*Salinispora tropica*) | β-Lactone-γ-lactam, hybrid PKS-NRP | Proteasome inhibitor; competitor defense | [5] |
| Halogenated diterpenes (Dictyotaceae) | Terpene-isoprenoid pathway | Defense against herbivores; antifouling | [25] |
| Halogenated furanones (*Delisea pulchra*) | Polyketide-terpene with halogenation | Block bacterial quorum sensing; antifouling | [26] |
| Hymenidin/debromo-carteramine A (sponges) | Bromopyrrole alkaloids from tryptophan | Defense against predators and microbes | [27] |
| Glycoglycerolipids (hydrolithon) | Lipid glycosylation | Induce coral settlement and reef recovery | [28] |

**Table 1.**
*Marine secondary metabolites, metabolic pathways, and ecological implications.*

### 2.2 Marine sources of bioactive metabolites

Marine biodiversity is the basis for the discovery of new bioactive metabolites. Marine microorganisms, particularly actinomycetes and cyanobacteria, are prolific producers of secondary metabolites (**Figure 1**). Actinomycetes isolated from sponges or sediments have yielded new macrolides and polyketides, while cyanobacteria generate cytotoxic peptides such as cryptophycins [31]. Marine invertebrates are equally important. Sponges produce various alkaloids and terpenes with therapeutic potential [32], while tunicates provide compounds such as trabectedin, now used clinically [33]. Macroalgae are also renewable reservoirs: brown algae contain phlorotannins and fucoidan, red algae provide carrageenans, and green algae produce ulvans with antioxidant properties [34].

### 2.3 Challenges associated with marine bioprospecting

Despite this promise, marine bioprospecting faces major challenges. Many metabolites have complex polycyclic structures and unusual substituents, which require advanced techniques such as high-resolution mass spectrometry and NMR to be elucidated [34]. Furthermore, symbiotic microbiomes can generate the same metabolites as their hosts, complicating traceability [35]. Traditional discovery strategies are laborious and costly. Recent advances integrate mass spectrometry with bioinformatic tools such as GNPS molecular networks and in silico annotation, supported by artificial intelligence, to accelerate identification and avoid rediscovery [12]. The structural diversity of marine metabolites therefore makes their study both challenging and highly promising.

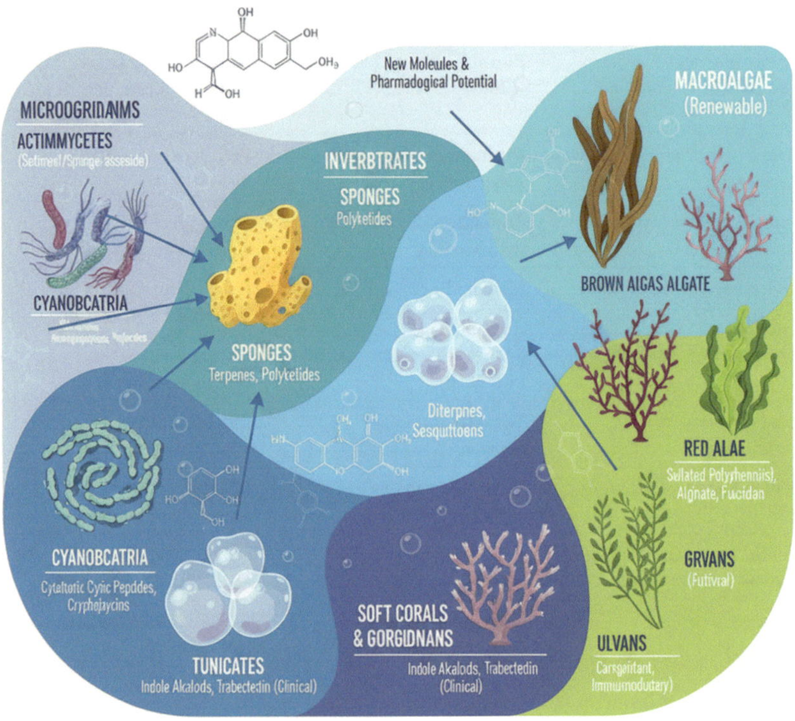


**Figure 1.**
*Marine biodiversity as a valuable source of novel bioactive metabolites with therapeutic and pharmacological potential (LLM: AI-generated image "ChatGPT").*

## 3. Analytical methodology centered on HPLC-MS

### 3.1 HPLC-MS (or LC-MS): Core of the analysis

#### *3.1.1 General principle*

High-performance liquid chromatography coupled with mass spectrometry (LC-MS) is the gold standard technique for studying marine metabolites due to its versatility and sensitivity. Separation is based on the interaction between the analytes and the stationary phase of the column. Reverse-phase C18 columns (RP-C18) are most commonly used for semi-polar molecules, while HILIC (hydrophilic interaction liquid chromatography) columns are better at retaining highly polar compounds, such as certain polysaccharides or amino metabolites [36]. Optimization of elution gradients (water/acetonitrile or water/methanol mixture, often acidified with formic acid) is essential for effective resolution.

HPLC-MS couples chromatographic separation with mass spectrometry for metabolite analysis [37]. After sample injection, the mobile phase carries compounds under high pressure through a column, where analytes are separated based on interactions with the stationary phase (**Figure 2**). The eluted molecules are ionized in the MS source, and the resulting ions are distinguished by mass-to-charge ratio (m/z), detected, and processed into spectra for identification and quantification.

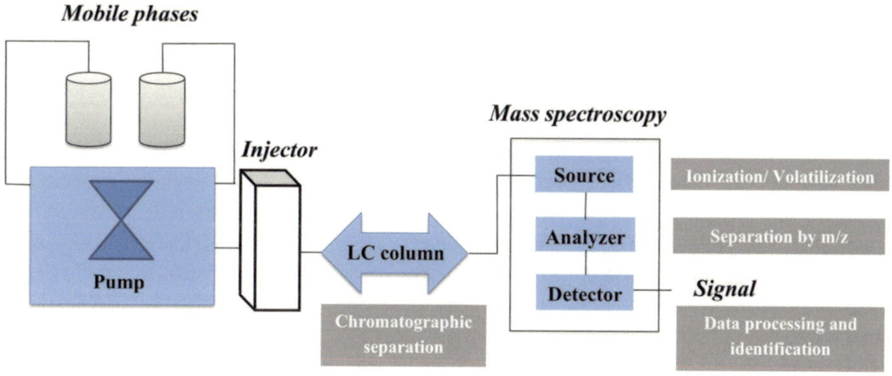


**Figure 2.**
*Principle schematic of an HPLC-MS system.*

### *3.1.1.1 Chromatographic separation*

Chromatographic separation is essential in metabolomics, as it isolates metabolites based on their chemical properties. Separation depends on analyte interactions with the stationary phase and on the composition of the mobile phase, including solvent type (acetonitrile and methanol), pH, buffers, and gradients. RP-C18 columns are the most common, suitable for semi-polar metabolites such as peptides, polyphenols, alkaloids, or pigments, though less effective for highly polar compounds [38]. For these, HILIC columns are preferred, offering strong retention of amino acids, sugars, nucleosides, and ionic metabolites [36]. In addition, ion exchange, normal phase, or mixed-phase supports are used for specific metabolite classes.

### *3.1.1.2 Ionization*

In marine metabolomics, two main atmospheric pressure ionization techniques are widely used. Electrospray ionization (ESI) is a soft method ideal for thermolabile, polar, and semi-polar metabolites such as peptides, glycosides, sulfated polysaccharides, and halogenated alkaloids [39]. It produces molecular or quasi-molecular ions ($[M + H]^+$, $[M - H]^-$, and adducts) but is sensitive to matrix effects, requiring optimization of solvents, flow, and additives. Atmospheric pressure chemical ionization (APCI) is more suited to less polar molecules (lipids, sterols, and pigments) [37], less affected by ion suppression, and often combined with ESI to broaden metabolome coverage.

### *3.1.1.3 Mass analyzers*

Once the ions have been generated, their separation and detection rely on mass analyzers with various characteristics (**Table 2**).

### *3.1.1.4 Data processing and identification*

The analysis of marine metabolomic data relies on MS and MS/MS spectra, which are compared with specialized databases such as GNPS, MassBank, and MarinLit, the latter being highly relevant for marine compounds. High-resolution analyzers, including Orbitrap and Q-TOF, enable sensitive detection and annotation of unknown molecules [41]. Typical workflows involve chromatographic peak detection

| Analyzer | Principal advantages | Limitations/points to consider |
|---|---|---|
| Single quadrupole/ triple quadrupole (Q/ QqQ) | Excellent for targeted quantification in MRM mode, offering high selectivity and sensitive detection. | Requires standards to optimize transitions, less useful for discovering new compounds. |
| Time of flight (TOF) | Large mass range, high acquisition speed, useful for non-targeted studies/global profiling. | Resolution lower than Orbitrap or FT-ICR, with only moderate mass accuracy. |
| Orbitrap [40]. | Delivers ultra-high resolution and excellent mass accuracy, enabling structural elucidation and precise molecular formula determination. | High cost and complexity, requiring precise calibration and longer analyses to ensure accuracy. |
| Hybrid analyzers (Q-TOF, Q-Orbitrap, etc.) [41] | Combines quadrupole selectivity with TOF/Orbitrap resolution, enhancing MS/ MS fragmentation, identification, and quantification. | Increased complexity, more parameters to optimize, more demanding data management. |

**Table 2.**
*Summary of mass analyzer characteristics.*

and alignment, extraction of precursor and fragment ions, filtering of adducts and isotopes, and the use of bioinformatic tools such as MZmine 2, XCMS, or MS-Dial for untargeted analyses [42].

### *3.1.2 Practical applications*

LC-MS has become indispensable in marine bioprospecting due to its versatility and sensitivity. It is used for global metabolomic profiling, providing detailed chemical maps of algae, sponges, and symbiotic bacteria, and for assessing variations linked to environmental conditions [43]. The technique also supports dereplication by comparing exact masses and fragmentation patterns with databases; for instance, trabectedin was rapidly identified in several tunicates using LC-HRMS [44]. Furthermore, MS/MS fragmentation facilitates structural elucidation of new metabolites, while its ability to simultaneously quantify diverse families of compounds, such as phlorotannins and carotenoids in algae, gives an integrated view of marine chemical diversity [45].

### *3.1.3 Advantages and disadvantages*

The main strength of LC-MS lies in its sensitivity, which allows detection at levels ranging from femtograms to nanograms [46], and in its ability to directly analyze complex marine matrices without extensive purification [36]. Its integration with untargeted metabolomics, molecular networks, and in silico annotation enhances compound discovery [47] Nevertheless, this technique faces challenges such as ion suppression by salts and pigments [34], its limited suitability for volatile metabolites that are better handled by GC-MS [48], and its high costs requiring frequent calibrations and advanced bioinformatics support [49].

## 3.2 GC-MS: For volatile compounds

### *3.2.1 General principle*

Gas chromatography coupled with mass spectrometry (GC-MS) is based on the separation of volatile and thermostable compounds in a capillary column before their detection by mass spectrometry. The molecules must be volatile and generally derivatized (methylation and silylation) to improve their thermal stability [48]. Ionization is most often performed by electron impact (EI), generating reproducible fragmentation spectra that facilitate comparison with mass libraries (NIST, Wiley) [50].

#### *3.2.2 Practical applications*

In marine metabolomics, GC-MS is widely used to characterize the lipid profiles of microalgae and macroalgae, particularly polyunsaturated fatty acids (EPA and DHA) after fatty acid methylation [51]. It also allows the identification of halogenated terpenes in corals and algae, compounds of ecological and pharmacological interest, and the detection of volatile organic compounds (VOCs) from marine bacteria, thus providing insight into symbiotic chemical communication [52].

#### *3.2.3 Advantages and disadvantages*

Gas chromatography coupled with mass spectrometry (GC-MS) is a reference method in marine metabolomics, valued for its robustness, high chromatographic resolution, and reproducibility. Its use of standardized electron impact spectra allows efficient compound identification through large spectral libraries [53]. GC-MS is especially suitable for analyzing small stable molecules and volatile lipids. Nonetheless, it has limitations: it is restricted to volatile and thermostable compounds, often requiring laborious derivatization, and proves less effective for polar or high-mass metabolites frequently found in algae and sponges [34].

### 3.3 MALDI-TOF-MS: Biomarkers and macromolecules

#### *3.3.1 General principle*

Matrix-assisted laser desorption/ionization time-of-flight mass spectrometry (MALDI-TOF-MS) is based on the gentle ionization of samples mixed with a laser-absorbing organic matrix. This technique generates intact ions, even for fragile macromolecules (proteins, polysaccharides, and cyclic peptides), which are then separated according to their time of flight [54].

#### *3.3.2 Practical applications*

In marine bioprospecting, MALDI-TOF-MS is a valuable tool for investigating chemical and microbial diversity. It is particularly effective in identifying bioactive peptides from cyanobacteria, supporting the discovery of new therapeutic candidates [31]. The technique is also applied to the rapid analysis of sulfated glycans in brown algae, compounds of major biological and biotechnological interest [29]. Furthermore, MALDI-TOF-MS is widely used for microbial screening via protein profiling, enabling the identification of sponge-associated bacteria and offering key insights into marine microbial diversity and metabolic potential [55].

#### *3.3.3 Advantages and disadvantages*

MALDI-TOF-MS is emerging as a fast and practical technique for biomolecular analysis, relying on a straightforward preparation step in which the sample is mixed with a matrix. This approach helps preserve fragile macromolecules by minimizing fragmentation and is widely used as a screening tool for microbial identification [55]. Despite these advantages, it has notable limitations: its resolution and mass accuracy are lower than those of high-resolution instruments such as the Orbitrap. Moreover, the results are strongly influenced by matrix selection, due to ion suppression effects, and the method remains poorly suited for complex quantitative metabolomic studies [56].

### 3.4 HR-MS/MS/MS (Orbitrap, Q-TOF)

#### *3.4.1 General principle*

High-resolution instruments (HR-MS) such as the Orbitrap or the quadrupole time-of-flight (Q-TOF) allow accurate measurement of molecular masses with sub-ppm precision [40]. Coupled with tandem spectrometry (MS/MS), they provide fragmentation spectra rich in structural information via techniques such as collision-induced dissociation (CID) or higher-energy collisional dissociation (HCD) [57].

#### *3.4.2 Practical applications*

In marine chemistry, high-resolution tandem mass spectrometry (HR-MS/MS) is a key technique for investigating specialized metabolites. It is widely applied to the structural elucidation of complex polyketides from sponges, compounds with significant pharmacological potential. HR-MS/MS also accelerates dereplication by comparing MS/MS spectra to databases such as GNPS or MarinLit, allowing clear distinction between known and novel molecules [42]. Beyond this, fragmentation-guided strategies reveal diagnostic signatures that support the identification of new metabolite families, including peptides and halogenated alkaloids, underscoring its central role in marine bioprospecting [14].

#### *3.4.3 Advantages and disadvantages*

High-resolution mass spectrometry (HR-MS/MS), using Orbitrap or Q-TOF, is a powerful tool in marine chemistry due to its precision and ability to resolve closely spaced masses. It enables the simultaneous detection of hundreds of metabolites within complex extracts. Fragmentation data are well adapted to bioinformatic platforms such as GNPS, SIRIUS, or MS-FINDER, enhancing metabolomic characterization [49]. Despite its advantages, HR-MS/MS has limitations: high cost, specialized maintenance, and data complexity requiring advanced processing. In addition, co-elution in marine samples complicates signal deconvolution, reducing interpretive reliability [58].

## 4. Treatment of marine samples

The processing of marine samples is a crucial stage in the study and characterization of bioactive metabolites. The complexity of marine biological matrices (algae, sponges, symbiotic bacteria, tunicates, etc.) and the structural diversity of the compounds require appropriate extraction and purification processes. Challenges include high salinity, the presence of interfering pigments, and the often-low concentration of metabolites.

## 4.1 Extraction

### *4.1.1 General principle*

Extraction is a central step in marine chemistry, enabling the transfer of metabolites from biological matrices into appropriate solvent phases. The efficiency and selectivity of this process depend on solvent polarity relative to the target compounds. Methanol (MeOH) and hydroalcoholic mixtures are generally used for polar or semi-polar metabolites such as polysaccharides, phenols, and some alkaloids [59]. In contrast, chloroform ($CHCl_3$) is better suited for hydrophobic compounds, including lipids, terpenes, and carotenoids [60]. Ethyl acetate (EtOAc), with intermediate polarity, is widely employed for extracting polyketides and certain bioactive peptides [61].

### *4.1.2 Techniques*

Various extraction techniques are used in marine chemistry to effectively isolate metabolites of interest depending on the nature of the matrix and the polarity of the compounds. Soxhlet extraction is a classic method of continuous reflux extraction, particularly suitable for dry samples such as macroalgae or sediments [30]. It has been used in particular to recover phlorotannins from brown algae. Ultrasound-assisted extraction (UAE) is based on ultrasonic cavitation, which increases cell permeability and facilitates the release of metabolites. This rapid and economical approach is commonly applied to carotenoid-rich microalgae [62]. Pressure extraction (PLE or ASE) combines high temperatures (50–200°C) and high pressure (up to 200 bar) to accelerate extraction and is often used for marine polysaccharides and lipids [63]. Finally, supercritical fluid extraction (SFE) uses supercritical $CO_2$, sometimes modified with ethanol, to selectively extract lipids such as DHA and EPA in an environmentally friendly process [64]. These methods offer significant flexibility for marine bioprospecting.

### *4.1.3 Practical applications*

Marine chemistry applies diverse extraction strategies depending on metabolite type and matrix. Sulfated polysaccharides and polyphenols from *Fucus vesiculosus* are efficiently recovered using MeOH/$H_2O$ mixtures [29]. Halogenated alkaloids produced by tropical sponges are usually isolated with $CHCl_3$/MeOH solvent systems [32]. Long-chain polyunsaturated fatty acids (EPA and DHA) from microalgae such as *Nannochloropsis* and *Isochrysis* are extracted via supercritical $CO_2$, a cleaner and sustainable method. These techniques demonstrate the adaptability of solvents and processes to optimize yields across metabolite classes. Nonetheless, challenges remain, including the co-extraction of undesirable compounds, lengthy procedures, and high solvent requirements.

## 4.2 Purification

### *4.2.1 Principe general*

Raw extracts from marine organisms generally contain a wide variety of metabolites, often several hundred, which complicates their direct analysis [65].

In order to reduce this complexity and avoid interference that could distort the interpretation of the results, a purification step is essential. This step removes some of the unwanted compounds and concentrates the molecules of interest, thereby improving the sensitivity and reliability of subsequent LC-MS/MS analyses [65].

#### *4.2.2 Common techniques*

Several complementary purification methods are widely applied in marine natural product chemistry. Solid-phase extraction (SPE), using C18 or ion exchange cartridges, is one of the most common techniques. It is particularly effective for removing salts and sugars abundant in marine matrices while concentrating semi-polar metabolites before LC-MS analysis [64]. Flash chromatography, performed on silica or resin columns, enables rapid fractionation of complex extracts, especially sponge-derived samples rich in terpenes and lipophilic metabolites [66]. For advanced purification, semi-preparative HPLC allows isolation of pure metabolites in milligram quantities for subsequent structural studies, notably by NMR. A well-known case is trabectedin, originally isolated from tunicates, which became an approved anticancer drug [67]. Finally, bioassay-guided fractionation couples separation and biological testing, linking observed activities to specific metabolites identified by LC-MS [68].

#### *4.2.3 Practical applications*

Purification techniques play a central role in marine natural product chemistry, offering practical solutions for improving metabolite analysis. SPE-C18 cartridges are widely used to purify fucoidans from brown algae by eliminating salts and sugars, thereby enhancing the quality of mass spectrometry spectra [29]. Flash chromatography is effective for rapidly isolating lipophilic metabolites, such as the bioactive diterpenes from *Sinularia* corals, often investigated for anticancer and anti-inflammatory activities [66]. For rare or complex molecules present in trace amounts, semi-preparative HPLC is indispensable. It has enabled the fractionation of cyanobacterial cyclic peptides like cryptophycins, prior to structural elucidation by NMR and evaluation of cytotoxic activity [31]. These methods improve spectral clarity and detection of minor metabolites but remain time-consuming, technically demanding, and may lead to the loss of unstable compounds.

### 4.3 Challenges specific to marine samples

#### *4.3.1 High salinity*

A major challenge in marine metabolomics arises from the high salt content (NaCl, $MgSO_4$, and KCl) of marine extracts. Excess salts cause ion suppression in ESI-MS, lowering sensitivity and masking minor metabolites [69]. They also induce multiple adducts ($Na^+$ and $K^+$), complicating spectra interpretation and obscuring molecular ions. To reduce these effects, desalting methods such as dialysis, ion exchange, and ultrafiltration are applied. The most common is SPE-C18, which removes salts while concentrating semi-polar metabolites. Despite being time-consuming, these steps are essential for reliable LC-MS analyses of marine samples [46].

#### *4.3.2 Pigments and humic compounds*

Beyond salinity, pigments and humic substances are major obstacles in marine metabolite analysis. Macroalgae contain abundant chlorophylls and carotenoids, while sponges and sediments often accumulate humic matter from organic degradation. These compounds disrupt detection by absorbing light and generating background noise, which can mask metabolite ions and reduce MS sensitivity [70]. To minimize interference, selective extraction with solvents such as hexane removes non-polar pigments, while preparative chromatography separates cleaner fractions. Such purification improves MS spectra, facilitates metabolite identification and quantification, and reduces ion suppression, ensuring more reliable analyses.

#### *4.3.3 Low metabolite concentrations*

A key challenge in marine metabolomics is the very low abundance of many metabolites, often detected at nanogram-per-gram levels (ng/g). This scarcity complicates detection and may prevent complete characterization of an organism's chemical profile. Minor metabolites, though biologically significant, can be masked by major compounds or remain undetected in standard LC-MS analyses. To address this, extracts are typically preconcentrated through vacuum evaporation, freeze-drying, or solid-phase extraction (SPE), which enrich target metabolites while removing impurities. The use of high-resolution instruments such as Orbitrap or Q-TOF is also recommended, as they combine high sensitivity with superior mass accuracy, enabling the annotation of metabolites at trace levels [58]. Together, these strategies ensure reliable and reproducible analyses despite low metabolite abundance.

## 5. Identification and annotation of metabolites

The identification of marine metabolites is one of the most complex steps in natural product chemistry. It relies on a combination of spectrometric analyses, bioinformatic tools and complementary techniques (NMR, IR, and UV-Vis).

Mass spectrometry (MS) plays a central role in the identification of marine metabolites, providing accurate molecular mass and key structural information. Compounds are generally detected as quasi-molecular ions such as $[M + H]^+$ or $[M + Na]^+$ in positive mode and $[M - H]^-$ in negative mode, enabling precise mass determination and preliminary annotation. Isotopic analysis is particularly valuable for halogenated compounds, which are abundant in marine organisms: bromine produces a 1:1 doublet and chlorine a 3:1 ratio, allowing rapid recognition of halogenated metabolites from sponges or algae [53]. MS/MS fragmentation offers additional structural clues through characteristic neutral losses (e.g., 162 Da for glucose, 80 Da for sulfate, and 42 Da for acetyl), which guide structure elucidation before confirmation by NMR, IR, or UV-Vis spectroscopy [71].

Applications of MS in marine chemistry include the identification of halogenated alkaloids such as fascaplysine and aeroplysinin from sponges, characterized by their isotopic signatures, and sulfated glycosides in ascidians, revealed through successive losses of $SO_3$ and sugar residues. High-resolution mass spectrometry (HR-MS) using Orbitrap or Q-TOF analyzers enables rapid detection of polyhalogenated polyketides directly in crude extracts, facilitating dereplication and expanding the exploration of marine chemodiversity [72].

Bioinformatics has further transformed metabolite annotation. Databases such as MarinLit (over 35,000 entries) and the Seaweed Metabolite Database (www.swmd.co.in) focus on marine compounds [73], while platforms like GNPS, MassBank, and NPAtlas allow automated spectrum comparisons. Tools including GNPS Molecular Networking (https://gnps.ucsd.edu/), MolNetEnhancer, MZmine, MS-DIAL, and SIRIUS streamline data processing and structural prediction [53]. Together, these approaches represent a cornerstone of modern marine metabolomics, accelerating the discovery and characterization of novel bioactive metabolites [42, 49, 67].

## 6. Discovery approaches via LC-MS

LC-MS-based discovery strategies have revolutionized marine metabolomics by enabling both targeted and untargeted exploration of chemical diversity. Targeted LC-MS screening, often performed in Multiple Reaction Monitoring (MRM) mode, focuses on predefined metabolites or families, such as microcystins in cyanobacteria and sulfated polysaccharides like fucoidans in algae, allowing precise and sensitive quantification in complex matrices [31]. In contrast, untargeted screening records all ionic signals without prior assumptions, providing comprehensive metabolomic profiles that reveal interspecies or environmental variations, as observed between tropical and temperate algae or across algal seasonal cycles [36]. This approach is also essential for discovering novel metabolites, including previously unknown polyketides from marine sponges. Combined with molecular networking and bioinformatics tools such as GNPS, LC-MS/MS facilitates the annotation of known compound families while uncovering new chemical entities, thereby accelerating chemodiversity- and bioactivity-guided discovery of marine natural products (**Table 3**) [74].

| Approach | Example | Application results | Tools |
|---|---|---|---|
| LC-HRMS and MS/MS | *Discodermia* sponge | Discovery of anticancer macrolides and identification of several novel structures opened the way to pharmacological studies on their cytotoxic properties | LC-HRMS, MS/MS fragmentation |
| LC-HRMS and MS/MS | Cyanobacteria-producing depsipeptides | Identification of new families of depsipeptides highlighted their high structural and biological diversity | LC-HRMS, MS/MS fragmentation |
| LC-HRMS/MS and Bioinformatics | Global metabolite analysis | Construction of molecular networks linking metabolites according to spectral similarities revealed known chemical families and non-annotated nodes suggesting new entities | GNPS |
| Integrated strategy | Exploration of marine bioactive metabolites | Acceleration of the discovery of new chemical entities and bioactive metabolites in bioactivity- or chemodiversity-guided screening programs | LC-HRMS/MS and GNPS |

**Table 3.**
*LC-MS/MS and molecular networking approaches for the discovery of novel marine bioactive metabolites.*

## 7. Quantification and dynamic analysis

Quantification of marine metabolites by LC-MS is fundamental for linking chemical composition to ecological or pharmacological functions. It relies on calibration curves correlating signal intensity with concentration, using either external standards or isotopically labeled internal standards to correct for matrix effects in complex marine extracts. High-resolution LC-HRMS achieves ng/mL detection limits, enabling precise quantification of trace compounds [75]. This technique allows accurate measurement of metabolites such as trabectedin in sea squirts or phlorotannins and fucoidans in brown algae, revealing how environmental factors influence biosynthesis [68]. Dynamic LC-MS analysis further tracks metabolite production and transformation across time and conditions, from microbial biosynthesis (e.g., salinosporamide A) to pharmacokinetics of marine-derived drugs. These applications provide valuable insights into metabolite dynamics and biological functions, though limited isotopic standards and matrix interferences still challenge quantitative accuracy [76].

## 8. Illustration of qualitative analysis results

### 8.1 Analysis of an extract by LC-MS (ESI⁺)

LC-MS analysis in positive ESI mode revealed 21 chromatographic peaks between 0.1 and 9.5 min of retention time (RT) (**Figure 3**). The first, at 0.1 min, corresponds to the internal calibrant, while the others, distributed across three zones (RT < 4 min, 4–7 min, and >8 min), highlight the extract's complexity. Among these, compound 18 (black arrow) displayed a strong, well-defined signal at m/z = 449.40 (**Figure 4**), Eluting at RT 6.6 min, it corresponds to the sodium adduct [M + Na]+, consistent with the theoretical mass for $C_{30}H_{50}ONa^+$ (449.38). The molecular formula $C_{30}H_{50}O$ suggests a pentacyclic triterpene. The $[M + H]^+$ ion at m/z = 427.4221 confirmed identification as lupeol, a bioactive triterpene with antioxidant, anti-inflammatory, and anticancer properties.

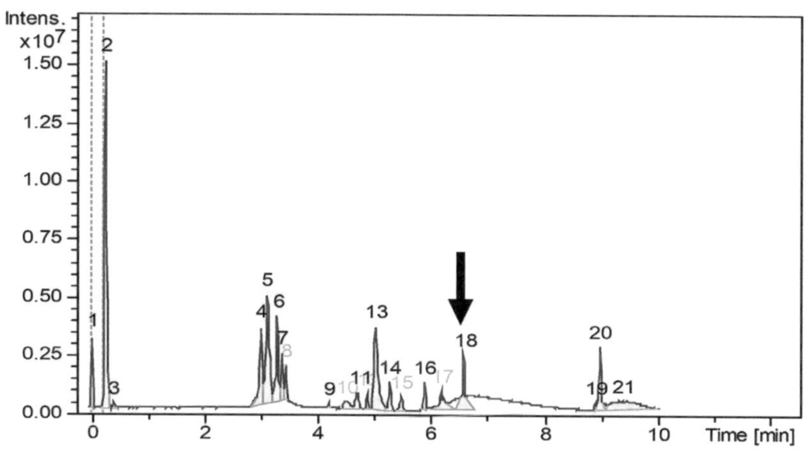


**Figure 3.**
*Chromatogram from LC-MS analysis.*

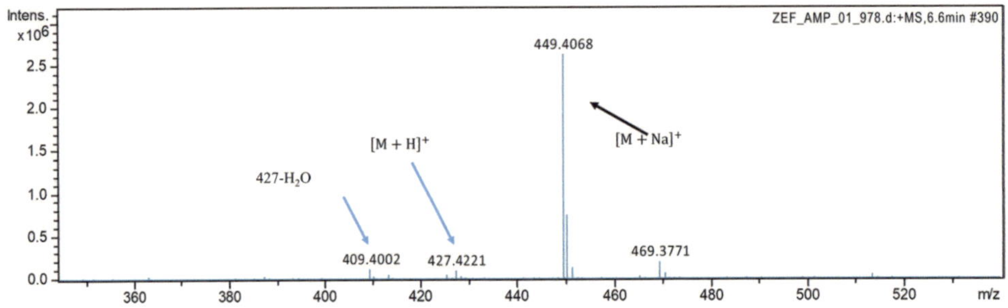


**Figure 4.**
*Positive mode HR-MS ESI mass spectrum of product 18.*

The mass spectrum reveals a neutral molecular mass near 426.41 uma, supported by the protonated ion at m/z = 427.4221 $[M + H]^+$ and the sodium adduct at m/z = 449.4068 $[M + Na]^+$, showing the expected 22 uma difference (Na-H). A key fragment at m/z = 409.4002 corresponds to $[M + H\text{-}H_2O]^+$, indicating dehydration linked to an alcohol group. Thus, the main diagnostic ions are m/z = 427.4221, 449.4068, and 409.4002. A weak peak at m/z = 469.3771 likely reflects a secondary adduct or solvent cluster. Further fragmentation involves C–C bond cleavages yielding stabilized tertiary carbocations, with possible 1,2-hydrogen or 1,2-methyl rearrangements, confirming fragmentation pathways typical of molecules with labile alcohol functions.

## 8.2 Analysis of a fraction by GC-MS$^n$ (EI)

The GC-MS chromatogram presented reveals the presence of four distinct compounds, identified by their retention times (**Figure 5**). Peaks A (26.5 min) and B (27.5 min) appear at low intensities, indicating minor and relatively volatile compounds. Peak C, at 34.0 min, clearly dominates the chromatogram in terms of intensity, reflecting the high abundance of this main compound in the sample. Finally, peak D (39.0 min) corresponds to a heavier or less volatile compound, detected in small quantities. This chromatographic profile shows effective separation and highlights the predominance of compound cover the other constituents.

GC-MS$^n$ (EI) analysis of a fraction revealed several minor constituents and one predominant metabolite, designated compound C. Its mass spectrum showed a molecular ion at m/z = 426.3, accompanied by characteristic fragment ions. The signal at m/z = 411.3 corresponds to the loss of a methyl group (M-15), while m/z = 393.3 reflects the loss of

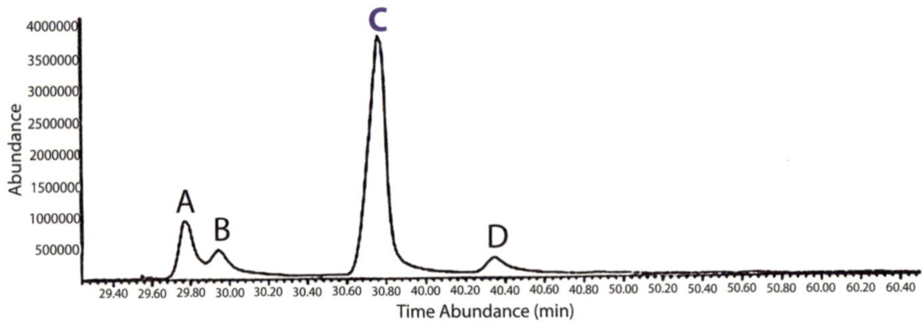


**Figure 5.**
*Chromatogram from GC-MS analysis.*

a water molecule (M-18). The intense base peak at m/z = 218.2, together with the ion at m/z = 207, is indicative of a double bond located at position 11–12 of a triterpene skeleton. These ions arise from a specific fragmentation pathway involving a Diels-Alder type rearrangement on the C ring. Furthermore, the ion at m/z = 189.1, derived from the dehydration of the m/z = 207 fragment, strongly suggests the presence of a hydroxyl group on one of the A or B rings, most likely at position 3, a common feature in triterpene biosynthesis. Collectively, these results point to an amyrin-type structure, though NMR is required to determine whether it corresponds to α- or β-amyrin (**Figure 6**).

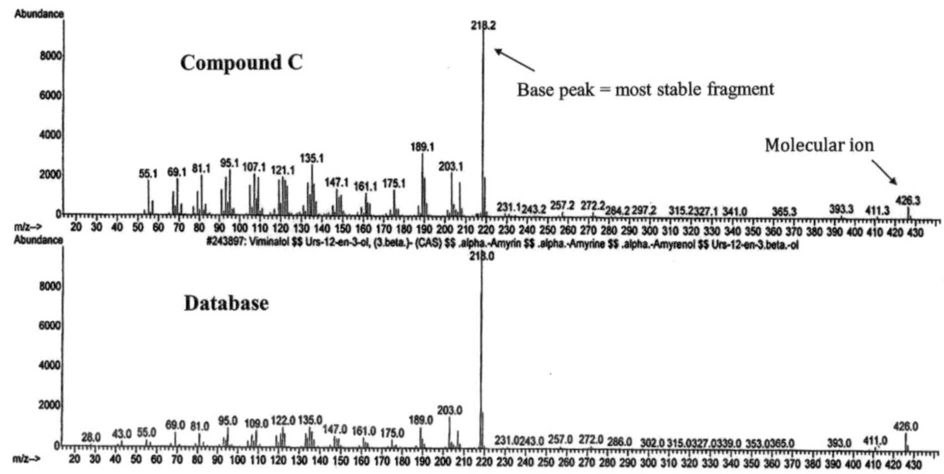


**Figure 6.**
*EI-MS[n] mass spectrum of compound C (top) and a reference product (bottom).*

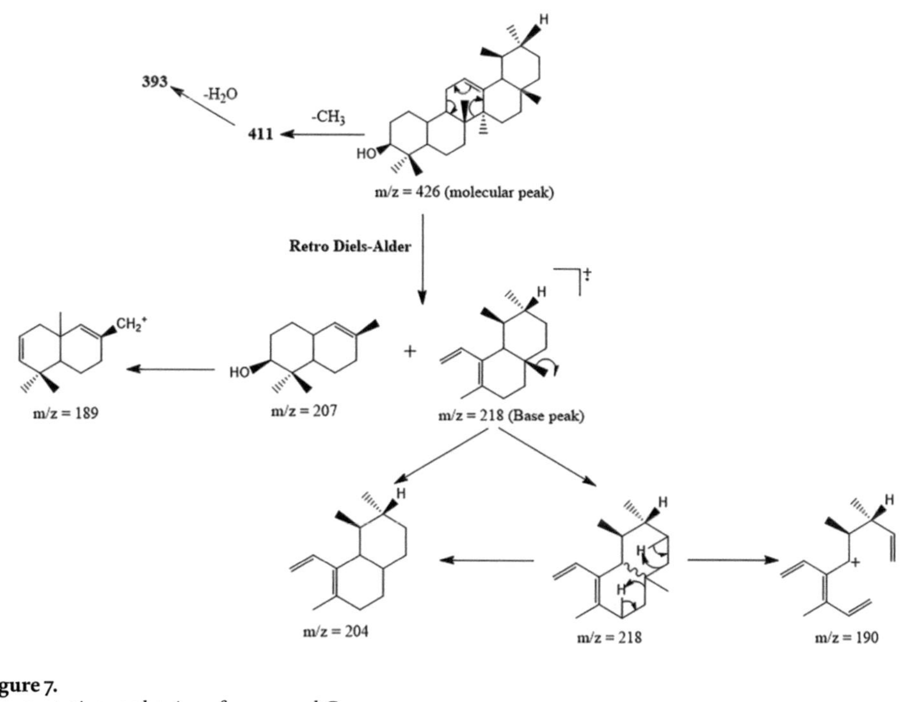


**Figure 7.**
*Fragmentation mechanism of compound C.*

The mass spectrum of compound C shows a molecular ion at m/z = 426, confirming its molecular weight. Initial fragmentation involves the loss of a methyl group (m/z = 411) and a combined elimination (m/z = 393). A prominent retro Diels-Alder reaction cleaves the polycyclic structure, producing complementary ions at m/z = 207 and the base peak m/z = 215, the most stable and intense fragment (**Figure 7**). Subsequent rearrangements and losses of small groups generate secondary ions at m/z = 204, 203, and finally a stabilized fragment at m/z = 190. The spectrum thus reflects progressive losses and a key cleavage yielding highly stable ions.

## 9. Current challenges and analytical perspectives

The study of marine bioactive metabolites using mass spectrometry has advanced considerably over the past two decades, thanks to high-resolution instruments (Orbitrap, Q-TOF) and advanced bioinformatics approaches. However, several methodological limitations remain, hindering the discovery and characterization of new molecules. The outlook is based on innovations in artificial intelligence, spatial imaging of metabolites, and sustainable biotechnology.

### 9.1 Limitations

Despite major advances in high-resolution mass spectrometry and bioinformatics tools, several obstacles remain in the study and characterization of marine metabolites. **Table 4** summarizes the main limitations encountered, their description, and their consequences for the discovery of new bioactive molecules.

### 9.2 Outlook for application

Future advances in marine metabolite chemistry depend on integrating emerging technologies to overcome current challenges. Artificial intelligence (AI) and machine learning greatly enhance mass spectra annotation. Models trained on large datasets can recognize complex fragmentation patterns and predict chemical classes, reducing reliance on limited libraries [77]. Tools such as CSI:FingerID and DeepMass enable

| Limit | Description | Consequences |
|---|---|---|
| Isolation of minor compounds | Marine metabolites are often present in very low concentrations, making their complete characterization difficult, even with LC-HRMS and multidimensional NMR [58]. | Limited access to new bioactive molecules; obstacle to drug candidate development. |
| Spectral redundancy and chemical complexity | Mass spectra frequently exhibit interference from multiple adducts, indistinguishable structural isomers, and background noise caused by salt- and pigment-rich marine matrices [69]. | Difficult automatic annotation, risk of false positives, less reliable untargeted analyses. |
| Limited databases | Despite databases like MarinLit and SeaweedMetaboliteDB, many marine spectra remain absent from public repositories such as GNPS or MassBank [2]. | Slow dereplication and incomplete recognition or annotation of known compounds. |

**Table 4.**
*Current limitations of mass spectrometry applied to marine metabolites.*

faster, more reliable annotation, opening the way to novel metabolite discovery. Molecular imaging (MSI) with LC-MS, MALDI-MSI, or DESI-MS allows spatial mapping of metabolites directly in tissues of algae, sponges, or cyanobacteria, linking chemistry to ecological functions. MicroLC-MS and nanoLC-MS, through miniaturization, improve sensitivity and reduce biomass and solvent use, making them ideal for rare extracts or metabolites at trace levels. Together, these approaches expand discovery potential while clarifying ecological and functional roles.

## 10. Conclusion

Mass spectrometry is a key tool for studying and exploiting marine bioactive metabolites. Thanks to its sensitivity and high resolution, it can identify, characterize, and quantify compounds present in trace amounts in complex marine matrices. Advances in LC-MS, GC-MS, MALDI-TOF, and HR-MS/MS techniques, combined with bioinformatics tools, have significantly accelerated the discovery of new molecules with pharmacological, cosmetic, and nutraceutical potential. However, challenges remain, related to the structural complexity of metabolites, their low abundance, and interference from salts or pigments. The integration of emerging technologies, such as artificial intelligence, metabolic imaging, and multi-omics approaches, opens up new perspectives for overcoming these limitations. By linking analytical innovation and sustainability, mass spectrometry promotes the rational exploitation of marine resources and offers unique opportunities for research and industry.

## Acknowledgements

The authors would like to express their sincere gratitude to the Faculty of Science at the Université d'Antananarivo (URGPGE, ISTE, PEI, LCSN/COB, and LCO-SM laboratories) for their support and access to facilities, as well as to the Université Bretagne Sud (LBCM laboratories) for their valuable collaboration and expertise, which greatly contributed to the success and quality of this study.

## Conflict of interest

The author declares no conflict of interest.

## List of abbreviations

| | |
|---|---|
| $[M + H]^+$ | protonated molecular ion |
| $[M + K]^+$ | potassium adduct ion |
| $[M + Na]^+$ | sodium adduct ion |
| $[M - H]^-$ | deprotonated molecular ion |
| AI | artificial intelligence |
| ASE | accelerated solvent extraction |
| CID-MS | collision induced dissociation-mass spectrometry |
| CSI:FingerID | competitive fragmentation modeling for metabolite identification |
| DeepMass | AI-based spectral annotation tool |

| | |
|---|---|
| DESI-MS | desorption electrospray ionization-mass spectrometry |
| DHA | docosahexaenoic acid |
| EI | electron ionization |
| EPA | eicosapentaenoic acid |
| ESI | electrospray ionization |
| FT-MS | Fourier transform mass spectrometry |
| GC-MS | gas chromatography-mass spectrometry |
| GNPS | Global Natural Products Social Molecular Networking |
| HPLC-MS | high-performance liquid chromatography-mass spectrometry |
| HR-MS or HRMS | high-resolution mass spectrometry |
| IR | infrared spectroscopy |
| IRMPD | infrared multiple photon dissociation |
| LC-HRMS | liquid chromatography-high-resolution mass spectrometry |
| LC-MS | liquid chromatography-mass spectrometry |
| LOD | limit of detection |
| LOQ | limit of quantification |
| MALDI-MSI | matrix-assisted laser desorption/ionization-mass spectrometry imaging |
| MALDI-TOF | matrix-assisted laser desorption/ionization-time of flight |
| MolNetEnhancer | molecular network enhancer |
| MRM | multiple reaction monitoring |
| MS | mass spectrometry |
| MS/MS | tandem mass spectrometry |
| MS-DIAL | Mass Spectrometry Data Independent AnaLysis |
| MSI | mass spectrometry imaging |
| MZmine | software for mass spectrometry data processing |
| NMR | nuclear magnetic resonance |
| NPAtlas | natural products atlas |
| NRPS | non-ribosomal peptide synthetase |
| OSMAC | one strain many compounds |
| PKS | polyketide synthase |
| PLE | pressurized liquid extraction |
| PUFAs | polyunsaturated fatty acids |
| Q-TOF | quadrupole time-of-flight mass spectrometer |
| SFE | supercritical fluid extraction |
| SIRIUS | software for structure prediction from tandem mass spectra |
| SPE | solid phase extraction |
| SWMD | seaweed metabolite database |
| UV-Vis | ultraviolet-visible spectroscopy |

## Author details

Elando Fréda Zamanileha[1*], Marnella Quetty Lema[2], Antonio Randriampahasoavana[3], Stanislas Fulgence Lahadson[4], Picardino Frienduc Vaonalamihanta[5], Andriamanarivosoa Rija Razafintsalama[1], Gilles Bedoux[6] and Pierre Hervé Ravelonandro[1]

1 Faculty of Sciences, Unité de Recherche en Génie de Procédés et Génie de l'Environnement (URGPGE), Université d'Antananarivo, Antananarivo, Madagascar

2 Faculty of Sciences, Engineering in Water Science and Technology (ISTE), Université d'Antananarivo, Antananarivo, Madagascar

3 Faculty of Sciences, Processes and Industrial Ecology (PEI), Université d'Antananarivo, Antananarivo, Madagascar

4 Faculty of Sciences, Laboratory of Natural Substances Chemistry and Biological Organic Chemistry (LCSN/COB), Université d'Antananarivo, Antananarivo, Madagascar

5 Faculty of Sciences, Laboratory of Organic Chemistry-Marine Substances (LCO-SM), Université d'Antananarivo, Antananarivo, Madagascar

6 Université Bretagne Sud, EMR CNRS, LBCM, IUEM, Vannes, France

*Address all correspondence to: elando.zamanileha@outlook.com and elando.zamanileha@urgpge-ua.mg

## References

[1] Leal MC, Puga J, Serôdio J, Gomes NCM, Calado R. Trends in the discovery of new marine natural products from invertebrates over the last two decades—Where and what are we bioprospecting? PLoS ONE. 2012;7:e30580. DOI: 10.1371/journal.pone.0030580

[2] Blunt JW, Copp BR, Keyzers RA, Munro MHG, Prinsep MR. Marine natural products. Natural Product Reports. 2012;**29**:144-222. DOI: 10.1039/C2NP00090C

[3] Blunt JW, Copp BR, Munro MHG, Northcote PT, Prinsep MR. Marine natural products. Natural Product Reports. 2010;**27**:165. DOI: 10.1039/b906091j

[4] Hu G-P, Yuan J, Sun L, She Z-G, Wu J-H, Lan X-J, et al. Statistical research on marine natural products based on data obtained between 1985 and 2008. Marine Drugs. 2011;**9**:514-525. DOI: 10.3390/md9040514

[5] Martins A, Vieira H, Gaspar H, Santos S. Marketed marine natural products in the pharmaceutical and cosmeceutical industries: Tips for success. Marine Drugs. 2014;**12**:1066-1101. DOI: 10.3390/md12021066

[6] Mayer AMS, Mayer VA, Swanson-Mungerson M, Pierce ML, Rodríguez AD, Nakamura F, et al. Marine pharmacology in 2019-2021: Marine compounds with antibacterial, antidiabetic, antifungal, anti-inflammatory, antiprotozoal, antituberculosis and antiviral activities; affecting the immune and nervous systems, and other miscellaneous mechanisms of action. Marine Drugs. 2024;**22**:309. DOI: 10.3390/md22070309

[7] Newman DJ, Cragg GM. Natural products as sources of new drugs over the nearly four decades from 01/1981 to 09/2019. Journal of Natural Products. 2020;**83**:770-803. DOI: 10.1021/acs.jnatprod.9b01285

[8] Stiger-Pouvreau V, Jégou C, Cérantola S, Guérard F, Lann KL. Chapter thirteen—Phlorotannins in Sargassaceae species from Brittany (France): Interesting molecules for ecophysiological and valorisation purposes. In: Bourgougnon N, editor. Advances in Botanical Research. Vol. 71. Amsterdam, Netherlands: Academic Press; 2014. pp. 379-411. DOI: 10.1016/B978-0-12-408062-1.00013-5

[9] Suleria HAR, Osborne S, Masci P, Gobe G. Marine-based nutraceuticals: An innovative trend in the food and supplement industries. Marine Drugs. 2015;**13**:6336-6351. DOI: 10.3390/md13106336

[10] Mehbub MF, Lei J, Franco C, Zhang W. Marine sponge derived natural products between 2001 and 2010: Trends and opportunities for discovery of bioactives. Marine Drugs. 2014;**12**:4539-4577. DOI: 10.3390/md12084539

[11] Le VH, Inai M, Williams RM, Kan T. Ecteinascidins. A review of the chemistry, biology and clinical utility of potent tetrahydroisoquinoline antitumor antibiotics. Natural Product Reports. 2015;**32**:328-347. DOI: 10.1039/c4np00051j

[12] Dührkop K, Nothias L-F, Fleischauer M, Reher R, Ludwig M, Hoffmann MA, et al. Systematic classification of unknown metabolites using high-resolution fragmentation mass spectra. Nature Biotechnology.

2021;**39**:462-471. DOI: 10.1038/s41587-020-0740-8

[13] Harvey AL, Edrada-Ebel R, Quinn RJ. The re-emergence of natural products for drug discovery in the genomics era. Nature Reviews. Drug Discovery. 2015;**14**:111-129. DOI: 10.1038/nrd4510

[14] Mohimani H, Gurevich A, Mikheenko A, Garg N, Nothias L-F, Ninomiya A, et al. Dereplication of peptidic natural products through database search of mass spectra. Nature Chemical Biology. 2017;**13**:30-37. DOI: 10.1038/nchembio.2219

[15] Wijffels RH, Barbosa MJ. An outlook on microalgal biofuels. Science. 2010;**329**:796-799. DOI: 10.1126/science.1189003

[16] Martins AP, Colepicolo P, Yokoya NS. Concise review on seaweed photosynthesis: From physiological bases to biotechnological applications. Journal of Photochemistry and Photobiology. 2023;**16**:100194. DOI: 10.1016/j.jpap.2023.100194

[17] Lorente A, Makowski K, Albericio F, Álvarez M. Bioactive marine polyketides as potential and promising drugs. Annals of Marine Biology and Research. 2014;**3**:1-3. DOI: 10.47739/1003

[18] Tian Z, Lu X-T, Jiang X, Tian J. Bryostatin-1: A promising compound for neurological disorders. Frontiers in Pharmacology. 2023;**14**:1187411. DOI: 10.3389/fphar.2023.1187411

[19] Ali MA, Khan AU, Ali A, Khaliq M, Khan N, Mujahid S, et al. Didemnins as marine-derived anticancer agents: Mechanistic insights and clinical potential. Medical Oncology. 2025;**42**:43. DOI: 10.1007/s12032-024-02594-0

[20] Zhidkov ME, Smirnova PA, Tryapkin OA, Kantemirov AV, Khudyakova YV, Malyarenko OS, et al. Total syntheses and preliminary biological evaluation of brominated fascaplysin and reticulatine alkaloids and their analogues. Marine Drugs. 2019;**17**:496. DOI: 10.3390/md17090496

[21] Sneed JM, Sharp KH, Ritchie KB, Paul VJ. The chemical cue tetrabromopyrrole from a biofilm bacterium induces settlement of multiple Caribbean corals. Proceedings of the Royal Society B: Biological Sciences. 2014;**281**:20133086. DOI: 10.1098/rspb.2013.3086

[22] Gullo VP, McAlpine J, Lam KS, Baker D, Petersen F. Drug discovery from natural products. Journal of Industrial Microbiology & Biotechnology. 2006;**33**:523-531. DOI: 10.1007/s10295-006-0107-2

[23] Cueto M, Jensen PR, Kauffman C, Fenical W, Lobkovsky E, Clardy J. Pestalone, a new antibiotic produced by a marine fungus in response to bacterial challenge. Journal of Natural Products. 2001;**64**:1444-1446. DOI: 10.1021/np0102713

[24] Kwon HC, Kauffman CA, Jensen PR, Fenical W. Marinomycins A–D, antitumor-antibiotics of a new structure class from a marine actinomycete of the recently discovered genus "marinispora". Journal of the American Chemical Society. 2006;**128**:16410-16410. DOI: 10.1021/ja0666497. [J. Am. Chem. Soc. 2006, 128, 1622-1632]

[25] Craft JD, Paul VJ, Sotka EE. Biogeographic and phylogenetic effects on feeding resistance of generalist herbivores toward plant chemical defenses. Ecology. 2013;**94**:18-24. DOI: 10.1890/11-0873.1

[26] Harder T, Campbell AH, Egan S, Steinberg PD. Chemical mediation of ternary interactions between marine Holobionts and their environment as exemplified by the red alga Delisea pulchra. Journal of Chemical Ecology. 2012;**38**:442-450. DOI: 10.1007/s10886-012-0119-5

[27] Haber M, Carbone M, Mollo E, Gavagnin M, Ilan M. Chemical defense against predators and bacterial fouling in the Mediterranean sponges *Axinella polypoides* and *A. verrucosa*. Marine Ecology Progress Series. 2011;**422**:113-122

[28] Tebben J, Motti CA, Siboni N, Tapiolas DM, Negri AP, Schupp PJ, et al. Chemical mediation of coral larval settlement by crustose coralline algae. Scientific Reports. 2015;**5**:10803. DOI: 10.1038/srep10803

[29] Ale MT, Mikkelsen JD, Meyer AS. Important determinants for fucoidan bioactivity: A critical review of structure-function relations and extraction methods for fucose-containing sulfated polysaccharides from brown seaweeds. Marine Drugs. 2011;**9**:2106-2130. DOI: 10.3390/md9102106

[30] Zamanileha EF, Burlot A-S, Latire T, Marty C, Douzenel P, Vandanjon L, et al. Biochemical composition and seasonal variations of the Madagascar algae *Eucheuma denticulatum* (Solieriaceae, Rhodophyta). Marine Drugs. 2025;**23**:30. DOI: 10.3390/md23010030

[31] Tan LT. Bioactive natural products from marine cyanobacteria for drug discovery. Phytochemistry. 2007;**68**:954-979. DOI: 10.1016/j.phytochem.2007.01.012

[32] Sipkema D, Franssen MCR, Osinga R, Tramper J, Wijffels RH. Marine sponges as pharmacy. Marine Biotechnology (New York, N.Y.). 2005;**7**:142-162. DOI: 10.1007/s10126-004-0405-5

[33] Mayer AMS, Glaser KB, Cuevas C, Jacobs RS, Kem W, Little RD, et al. The odyssey of marine pharmaceuticals: A current pipeline perspective. Trends in Pharmacological Sciences. 2010;**31**:255-265. DOI: 10.1016/j.tips.2010.02.005

[34] Gross H, König GM. Terpenoids from marine organisms: Unique structures and their pharmacological potential. Phytochemistry Reviews. 2006;**5**:115-141. DOI: 10.1007/s11101-005-5464-3

[35] Piel J. Metabolites from symbiotic bacteria. Natural Product Reports. 2009;**26**:338-362. DOI: 10.1039/b703499g

[36] Wolfender J-L, Marti G, Thomas A, Bertrand S. Current approaches and challenges for the metabolite profiling of complex natural extracts. Journal of Chromatography. A. 2015;**1382**:136-164. DOI: 10.1016/j.chroma.2014.10.091

[37] Ganzera M, Sturm S. Recent advances on HPLC/MS in medicinal plant analysis—An update covering 2011-2016. Journal of Pharmaceutical and Biomedical Analysis. 2018;**147**:211-233. DOI: 10.1016/j.jpba.2017.07.038

[38] Ruiz-Angel MJ, Pous-Torres S, Carda-Broch S, García-Alvarez-Coque MC. Performance of different C18 columns in reversed-phase liquid chromatography with hydro-organic and micellar-organic mobile phases. Journal of Chromatography A. 2014;**1344**:76-82. DOI: 10.1016/j.chroma.2014.04.011

[39] Wilm M. Principles of electrospray ionization. Molecular & Cellular Proteomics. 2011;**10**:M111.009407. DOI: 10.1074/mcp.M111.009407

[40] Scigelova M, Hornshaw M, Giannakopulos A, Makarov A. Fourier transform mass spectrometry. Molecular & Cellular Proteomics. 2011;**10**:M111.009431. DOI: 10.1074/mcp.M111.009431

[41] Li Y, He C, Zhang Y, Shi Q. Online LC-Orbitrap MS method for the rapid molecular characterization of dissolved organic matter. Rapid Communications in Mass Spectrometry. 2024;**38**:e9885. DOI: 10.1002/rcm.9885

[42] Wang M, Carver JJ, Phelan VV, Sanchez LM, Garg N, Peng Y, et al. Sharing and community curation of mass spectrometry data with Global Natural Products Social Molecular Networking. Nature Biotechnology. 2016;**34**:828-837. DOI: 10.1038/nbt.3597

[43] Dettmer K, Aronov PA, Hammock BD. Mass spectrometry-based metabolomics. Mass Spectrometry Reviews. 2007;**26**:51-78. DOI: 10.1002/mas.20108

[44] Yang JY, Sanchez LM, Rath CM, Liu X, Boudreau PD, Bruns N, et al. Molecular networking as a dereplication strategy. Journal of Natural Products. 2013;**76**:1686-1699. DOI: 10.1021/np400413s

[45] Balasubramaniam V, June Chelyn L, Vimala S, Mohd Fairulnizal MN, Brownlee IA, Amin I. Carotenoid composition and antioxidant potential of *Eucheuma denticulatum*, *Sargassum polycystum* and *Caulerpa lentillifera*. Heliyon. 2020;**6**:e04654. DOI: 10.1016/j.heliyon.2020.e04654

[46] Vuckovic D. Current trends and challenges in sample preparation for global metabolomics using liquid chromatography–mass spectrometry. Analytical and Bioanalytical Chemistry. 2012;**403**:1523-1548. DOI: 10.1007/s00216-012-6039-y

[47] Patti GJ, Yanes O, Siuzdak G. Metabolomics: The apogee of the omics trilogy. Nature Reviews. Molecular Cell Biology. 2012;**13**:263-269. DOI: 10.1038/nrm3314

[48] Sparkman OD, Penton Z, Kitson FG. Gas Chromatography and Mass Spectrometry: A Practical Guide. 2nd ed. Amsterdam, Netherlands: Academic Press; 2011

[49] Dührkop K, Fleischauer M, Ludwig M, Aksenov AA, Melnik AV, Meusel M, et al. SIRIUS 4: A rapid tool for turning tandem mass spectra into metabolite structure information. Nature Methods. 2019;**16**:299-302. DOI: 10.1038/s41592-019-0344-8

[50] Oberacher H, Whitley G, Berger B. Evaluation of the sensitivity of the 'Wiley registry of tandem mass spectral data, MSforID' with MS/MS data of the 'NIST/NIH/EPA mass spectral library'. Journal of Mass Spectrometry. 2013;**48**:487-496. DOI: 10.1002/jms.3184

[51] Banskota AH, Stefanova R, Sperker S, Lall SP, Craigie JS, Hafting JT, et al. Polar lipids from the marine macroalga *Palmaria palmata* inhibit lipopolysaccharide-induced nitric oxide production in RAW264.7 macrophage cells. Phytochemistry. 2014;**101**:101-108. DOI: 10.1016/j.phytochem.2014.02.004

[52] Schulz S, Dickschat JS, Kunze B, Wagner-Dobler I, Diestel R, Sasse F. Biological activity of volatiles from marine and terrestrial bacteria. Marine Drugs. 2010;**8**:2976-2987. DOI: 10.3390/md8122976

[53] McLafferty FW. Mass spectrometric analysis. Molecular rearrangements. Analytical Chemistry. 1959;**31**:82-87. DOI: 10.1021/ac60145a015

[54] Michael K, Franz H. Laser desorption ionization of proteins with molecular masses exceeding 10,000 Daltons. Analytical Chemistry. 1988;**60**:2299-2301. DOI: 10.1021/ac00171a028

[55] Seng P, Drancourt M, Gouriet F, La Scola B, Fournier P-E, Rolain JM, et al. Ongoing revolution in bacteriology: Routine identification of bacteria by matrix-assisted laser desorption ionization time-of-flight mass spectrometry. Clinical Infectious Diseases. 2009;**49**:543-551. DOI: 10.1086/600885

[56] Clark AE, Kaleta EJ, Arora A, Wolk DM. Matrix-assisted laser desorption ionization–time of flight mass spectrometry: A fundamental shift in the routine practice of clinical microbiology. Clinical Microbiology Reviews. 2013;**26**:547-603. DOI: 10.1128/cmr.00072-12

[57] Corinti D, Rotari L, Crestoni ME, Fornarini S, Oomens J, Berden G, et al. Protonated forms of naringenin and naringenin chalcone: Proteiform bioactive species elucidated by IRMPD spectroscopy, IMS, CID-MS, and computational approaches. Journal of Agricultural and Food Chemistry. 2023;**71**:4005-4015. DOI: 10.1021/acs.jafc.2c07453

[58] Wolfender J-L, Nuzillard J-M, Van Der Hooft JJJ, Renault J-H, Bertrand S. Accelerating metabolite identification in natural product research: Toward an ideal combination of liquid chromatography–high-resolution tandem mass spectrometry and NMR profiling, *in silico* databases, and chemometrics. Analytical Chemistry. 2019;**91**:704-742. DOI: 10.1021/acs.analchem.8b05112

[59] Salem M, Bernach M, Bajdzienko K, Giavalisco P. A simple fractionated extraction method for the comprehensive analysis of metabolites, lipids, and proteins from a single sample. Journal of Visualized Experiments. 2017;**124**:55802. DOI: 10.3791/55802

[60] Saini RK, Prasad P, Shang X, Keum Y-S. Advances in lipid extraction methods—A review. International Journal of Molecular Sciences. 2021;**22**:13643. DOI: 10.3390/ijms222413643

[61] Blunt JW, Copp BR, Munro MHG, Northcote PT, Prinsep MR. Marine natural products. Natural Product Reports. 2006;**23**:26. DOI: 10.1039/b502792f

[62] Chemat F, Rombaut N, Sicaire A-G, Meullemiestre A, Fabiano-Tixier A-S, Abert-Vian M. Ultrasound assisted extraction of food and natural products. Mechanisms, techniques, combinations, protocols and applications. A review. Ultrasonics Sonochemistry. 2017;**34**:540-560. DOI: 10.1016/j.ultsonch.2016.06.035

[63] Herrero M, Cifuentes A, Ibanez E. Sub- and supercritical fluid extraction of functional ingredients from different natural sources: Plants, food-by-products, algae and microalgae: A review. Food Chemistry. 2006;**98**:136-148. DOI: 10.1016/j.foodchem.2005.05.058

[64] Raghuvanshi R, Bharate SB. Preclinical and clinical studies on bryostatins, a class of marine-derived protein kinase C modulators: A mini-review. Current Topics in Medicinal Chemistry. 2020;**20**:1124-1135. DOI: 10.2174/1568026620666200325110444

[65] Popov RS, Ivanchina NV, Dmitrenok PS. Application of MS-based metabolomic approaches in analysis of starfish and sea cucumber bioactive compounds. Marine Drugs. 2022;**20**:320. DOI: 10.3390/md20050320

[66] König GM, Wright AD. Marine natural products research: Current directions and future potential. Planta Medica. 2007;**62**:193-211. DOI: 10.1055/s-2006-957861

[67] Rinehart KL, Holt TG, Fregeau NL, Stroh JG, Keifer PA, Sun F, et al. Ecteinascidins 729, 743, 745, 759A, 759B, and 770: Potent antitumor agents from the Caribbean tunicate Ecteinascidia turbinata. Journal of Organic Chemistry. 1990;**55**:4512-4515. DOI: 10.1021/jo00302a007

[68] Cragg GM, Newman DJ. Natural products: A continuing source of novel drug leads. Biochimica et Biophysica Acta (BBA) - General Subjects. 2013;**1830**:3670-3695. DOI: 10.1016/j.bbagen.2013.02.008

[69] Constantopoulos TL, Jackson GS, Enke CG. Effects of salt concentration on analyte response using electrospray ionization mass spectrometry. Journal of the American Society for Mass Spectrometry. 1999;**10**:625-634. DOI: 10.1016/S1044-0305(99)00031-8

[70] Pereira DM, Valentão P, Andrade PB. Marine natural pigments: Chemistry, distribution and analysis. Dyes and Pigments. 2014;**111**:124-134. DOI: 10.1016/j.dyepig.2014.06.011

[71] Gross JH. Mass Spectrometry. Cham: Springer International Publishing; 2017. DOI: 10.1007/978-3-319-54398-7

[72] Gribble GW. Naturally occurring organohalogen compounds—A comprehensive review. Progress in the Chemistry of Organic Natural Products. 2023;**121**:1-546. DOI: 10.1007/978-3-031-26629-4_1

[73] Davis GDJ, Vasanthi AHR. Seaweed metabolite database (SWMD): A database of natural compounds from marine algae. Bioinformation. 2011;**5**:361-364. DOI: 10.6026/97320630005361

[74] Nothias L-F, Petras D, Schmid R, Dührkop K, Rainer J, Sarvepalli A, et al. Feature-based molecular networking in the GNPS analysis environment. Nature Methods. 2020;**17**:905-908. DOI: 10.1038/s41592-020-0933-6

[75] Niessen WMA. Progress in liquid chromatography-mass spectrometry instrumentation and its impact on high-throughput screening. Journal of Chromatography. A. 2003;**1000**:413-436. DOI: 10.1016/s0021-9673(03)00506-5

[76] Lagas JS, Sparidans RW, van Waterschoot RAB, Wagenaar E, Beijnen JH, Schinkel AH. P-glycoprotein limits oral availability, brain penetration, and toxicity of an anionic drug, the antibiotic salinomycin. Antimicrobial Agents and Chemotherapy. 2008;**52**:1034-1039. DOI: 10.1128/AAC.01041-07

[77] Brouard C, Shen H, Dührkop K, d'Alché-Buc F, Böcker S, Rousu J. Fast metabolite identification with Input Output Kernel Regression. Bioinformatics. 2016;**32**:i28-i36. DOI: 10.1093/bioinformatics/btw246